Unsettling Science and Religion

RELIGION AND SCIENCE AS A CRITICAL DISCOURSE

Series Editors

Lisa Stenmark, San Jose State University
Whitney Bauman, Florida International University

Understanding religion and science as a critical discourse means building on theoretical issues and concerns to address social transformation, and issues of justice and global concerns. Contributions to this series will employ multiple perspectives upon the process of doing and thinking "science and religion" together, but ultimately see the relationship of religion and science as creating space for a kind of critical discourse. This might mean: exploring disagreements between two authoritative disciplines that challenge one another; incorporating critical theories and discourses (understood narrowly as the Frankfort School, and, more broadly as critical race theory and feminist, postcolonial, and queer approaches within the social sciences, natural sciences, or humanities); or a focus on voices from outside the dominant discourse, which in the case of this series means people from outside of the western academy. In each case, the goal is to shake up assumptions, challenge givens, and open up space for new questions and new perspectives so we can think about pressing problems in a more productive and inclusive way.

Unsettling Science and Religion: Contributions and Questions from Queer Studies, edited by Lisa Stenmark and Whitney Bauman

Unsettling Science and Religion

Contributions and Questions from Queer Studies

Edited by
Lisa Stenmark and
Whitney Bauman

LEXINGTON BOOKS
Lanham • Boulder • New York • London

Published by Lexington Books
An imprint of The Rowman & Littlefield Publishing Group, Inc.
4501 Forbes Boulevard, Suite 200, Lanham, Maryland 20706
www.rowman.com

Unit A, Whitacre Mews, 26-34 Stannary Street, London SE11 4AB

British Library Cataloguing in Publication Information Available

Library of Congress Cataloging-in-Publication Data

Library of Congress Control Number: 2018936925
ISBN: 978-1-4985-5641-5 (cloth : alk. paper)
ISBN: 978-1-4985-5642-2 (electronic)

∞™ The paper used in this publication meets the minimum requirements of American National Standard for Information Sciences—Permanence of Paper for Printed Library Materials, ANSI/NISO Z39.48-1992.

Printed in the United States of America

Contents

Acknowledgments

As with all books, this book would not have been possible without a whole lot of people, events, places, and things. First and foremost, we would like to thank the Institute for Religion in an Age of Science for allowing us to co-direct a conference in July of 2015 on Star Island, exploring what contributions, observations, insights, techniques, and challenges QT might make to our understandings of the relationship between religion and science, and the science and religion discourses (the SRD). This conference was both fruitful and exciting; not just the papers but the conversations—sometimes late into the night, under falling stars and fueled by wine—about what it would mean to "queer" the discourses of religion and science and, by extension, the relationship between them. These conversations were so invigorating that it was clear that we needed to bring them to a broader audience, and thus this volume was born. Many of the contributors in this volume were a part of that conference, while others were invited to contribute after the fact (several of whom had been invited to present, but were unable to attend). Clearly, this volume would not have been possible without this organization, and without the beautiful and unique environment of the Star Island retreat center.

Secondly, we would like to thank all of those who contributed to that conference and this volume (both those included in this volume and those not). Third, we would like to thank the American Academy of Religion. Without the space of engagement and interaction made possible by the annual meeting of the American Academy of Religion, the two editors would likely have not been able to get much of the work done on editing the volume, nor had the opportunity to meet our wonderful editor, Sarah Craig. Sarah has been encouraging all the way through in regards to this project and to the series of which it is a part. Finally, we would like to thank our spouses for their support and putting up with the hours we spent editing and discussing this project on Skype. We hope that this volume will be worth it for the reader and all those who have contributed to its emergence.

Introduction

Lisa Stenmark and Whitney Bauman

Although we anticipate that most readers will be at least somewhat familiar with the contours of both queer theory (QT) and the science and religion discourses (the SRD), it is the nature of interdisciplinary collections that few will have a deep knowledge of both, and for that reason we begin with a brief overview of each. While this will, by necessity, be somewhat cursory, it will provide the necessary framework for understanding what follows. We hope that those familiar with these discourses will also appreciate this overview. For those who want even more detail, the bibliography for each chapter is a good place to start, and for those beginners who want to know more about one or more of the areas this book explores, we have included a brief annotated bibliography that describes a few introductory texts. We hope the reader will find these features helpful.

QUEER THEORY

"Queer theory" is an umbrella term that refers to a series of intellectual currents that began to coalesce in the late 1980s to mid 1990s. QT challenges assumptions about what is perceived as given or natural—including such things as sex, gender, and race—and posits that they are instead cultural constructs. Intellectually, QT traces its roots to postmodernism and post-structuralism, particularly to the work of Jacques Derrida and Michel Foucault, and to Judith Butler's problematization of the modern concepts of identity. Indeed, many of the papers in this volume make reference to these thinkers. These intellectual ideas shaped, and were shaped by, political, artistic, scientific, and religious developments, as well as parallel discourses such as feminist theory, critical race theory, and postcolonial critiques which challenged a wide variety of orthodox understandings of the way things "are."

Postmodernism/Post-structuralism

Postmodernism is first and foremost a critique of fixed universals and of grand narratives. The compulsion towards universals is not new to Modernity.[1] What is new is that these universals are now linked to human knowledge—particularly, but not exclusively, knowledge produced by modern science—which is in turn intertwined with progress. To be "Modern" is to think of knowledge as independent of human culture or context, produced by scientific and rational thought, and inextricably linked to progress. In the grand narrative of Modernism (scientific) knowledge equals progress. Postmodernism is the attempt to dismantle the metanarrative of modernism, demonstrating that what we know is neither fixed nor certain, and to break the link between "knowledge" and "progress." Several chapters in this volume utilize these critiques; in particular, see the chapters by Philip Clayton and Kirianna Florez, Lisa Stenmark, and Whitney Bauman. One way to launch this critique is to deconstruct the language around these givens, explore the history of how these ideas developed, and reveal their contingent nature. No thinkers are as closely connected with this process as Jacques Derrida and Michel Foucault.

Jacques Derrida: Deconstructing Language

We think words name *things*, that their meaning is grounded in those things, and for this reason our words connect us to the "real" world. Jacques Derrida challenged this perception, arguing that the meaning of words is not based on correspondence to reality, but on context: its relationship to other words, the rules of grammar and historical context. Even if we are familiar with the word "pen" for example, out of context the meaning of the word itself is unclear: is it an enclosure or a writing instrument? An action or a thing? This ambiguity is why, when someone asks what a word means, we often ask to hear it in a sentence. Saying, "I am writing with a fountain pen" provides a context that clarifies the meaning of "pen" by establishing a relationship between "pen, "writing," and "fountain." That doesn't end it, of course, because in addition to the linguistic context, there is a historical context—the meaning of "pen in relation to "fountain" which has changed over time: A fountain pen was originally a pen with an internal supply of ink—it flowed like a fountain—as opposed to one that needed to be dipped. It has come to mean a pen with a nib. So absent historical context, it is still not clear what I am writing with. All of this multiplicity is present whenever we use language—any language—and we cannot control the meaning of language because we cannot control these multiple contexts.

The situation is exacerbated by an additional element, which might be thought of as the context of the human mind, in which we understand what something is in relation to what it is not. This is the somewhat intuitive observation that the meaning of "up" depends on understanding "down," and "hot" provides the context for understanding "cold" (and vice versa). It is easy to lose sight of way words and concepts become meaningful only in relation to other words and concepts—the meaning of words seem fixed and natural, particularly when there is a perceived connection to a physical phenomenon: "Hot is hot!" Another mundane example: the meaning of "chair" seems straightforward, and indeed the *Wikipedia* entry for "chair" provides a simple definition: "a piece of furniture with a raised surface" which is "commonly used to seat a single person." But this simple definition becomes complicated very quickly. In addition to the ambiguity of "single"—is it for any one person or only someone who is unmarried?—there are a vast multitude of things we can sit on—a sofa, a bench, the dock of the bay—so *Wikipedia* begins to make some distinctions: a chair is not a stool ("a chair without a back or arm rests"), a seat (a "permanently fixed chair"), a couch (an "upholstered, padded chair for more than one person"), and so on. What began as a description of "chair" has become a description of what a chair is not. But this shift goes largely unnoticed, and suddenly chair has become a given, a concept by which we define other things—stool, seat, couch—can be reliably understood, even though we are still defining chair. This sense of given-ness is reinforced by the physical existence of chairs—when pressed, I can point to a chair as though that settles its meaning—*Wikipedia* in fact uses an image of a chair to define "chair."

Chair is not a fixed category, however, and its meaning has been porous. The love seat, for example, was originally a chair which could accommodate the fashionably large skirts of the early 19th century. Changing fashions left it stranded—not quite a chair, not quite a sofa—until it became its own category: "a sofa designed for seating two persons." Chair is not an isolated example, all concepts are defined through exclusions and thus all categories are unstable because no system can be comprehensive; the context changes and there will always be things that exist outside the box. These remainders or supplements must either be ignored (and thus rendered invisible) or meanings must shift, perhaps new categories are created, by focusing on (and thus accentuating) some similarities, while simultaneously ignoring (and thus obscuring) some differences.

Derrida saw binary thinking as central to Western thought, along with the tendency to connect it to observed regularities in the natural world. These binaries are embedded in text and language, which in turn establish the boundaries of thought by which we are governed and by which we understand the world. There are many languages, and many ways of understanding the world, but we have no control over what language, and thus what worldview, we are born into—we write and think in a language we did not create and a logic that we do not fully control and that language and that logic control us. The best we can do is to understand the words that we use, and to open our worlds up by deconstructing them: disrupting binaries and making what is marginalized central, seeking out and exploiting the ambiguities and limits of language to demonstrate that the ways that concepts that appear fixed, given, or natural are in fact constructed. There are a number of chapters that engage in deconstructing various texts; see especially the chapters by Emilie Townes, Carol Wayne White, Julia Watts Belser, Zairong Xiang, Teresa Hornsby, and Fern Feldman.

One of the places that Derrida demonstrates the importance of this kind of deconstruction is in his discussion of race and racism. Naturalistic justifications of "race" involve disguising the ways that "race" is merely a series of overlapping similarities and differences which have been determined differently by different people(s) at different times. These lines are arbitrary because there is no essential difference that makes race anything other than a textual fiction. For Derrida, "there's no racism without language."[2] The critical race theories of Franz Fanon and others make similar claims about the cultural, linguistic, and historical relativity of "race." There are a number of chapters in this volume engaged in demonstrating that the concepts of race—and thus racism—are intertwined with religion (Wayne White) and science (Townes), and the way all three are intertwined in colonialism (Stenmark). In addition, Watts Belser's chapter deals with destabilizing both race and ableism.

Michel Foucault: Discourse and Power

While Jacques Derrida focused on texts, Michel Foucault was more interested in the discursive practices that caused a text to be produced.[3] For Foucault, a discursive practice describes the relationship between language, the production of knowledge, and power—he was particularly interested in the newly emerging human sciences of medicine, psychiatry, and the social sciences. To simplify: society and the institutions within it have a set of unwritten rules about who gets to produce knowledge, and what they produce is knowledge about other people. This sets up an unequal power relationship—those with the power to name have power over those so named—and those power relationships permeate society. The central mechanism for the exercise of discursive power is the distinction between "normal" and "abnormal," in which "normal" emerges as a series of binary oppositions with those things deemed "abnormal" or "deviant." As with "chair" above, "normal" only emerges in the process of identifying and defining "not normal"—e.g., insanity, criminality, and deviant sexuality. This process of defining through exclusion and normalization works as a mechanism for social control. First, because the norms of behavior established by these definitions reinforce particular ideologies: definitions of insanity enforced Enlightenment norms, while ideas about "normal" sexuality supported the need for a labor supply in emerging capitalist economies. Moreover, certain discursive practices establish relations of power in which the producers of knowledge—experts—exert power by defining who and what is deviant or abnormal, while those persons become the object of knowledge, excluded from the process of defining it, and rendered voiceless and powerless. This power relationship is illustrated by the clinic, in which a group of experts crowd around a patient who is an object of study, reduced to a set of symptoms.

The connection between the human sciences and social control leads Foucault to assert that within modern society, knowledge and power are inseparable—one cannot exercise power without producing knowledge, and the production of knowledge is only possible through the exercise of power. What is distinct, is that this kind of power is not external, it is internal, it is a new sort of power—discursive power—which has emerged in the modern era. Previously, power was understood in terms of the ability to control behavior—sometimes through violence. But, discursive power does not control *actions*, it controls *persons*—Foucault calls this bio-power. Bio-power is not external, because we internalize the judgments of experts, we accept what doctors tell us about our bodies, even when it conflicts with our own experience. But bio-power is not merely a matter of accepting the judgment of experts when it contradicts your judgment; the expert's judgment *becomes* your judgment, it becomes the way you see and experience the world. It is "naturalized" as the way things are. This type of power permeates society, and the techniques for controlling and normalizing behavior are found in a variety of institutions—schools, hospitals, prisons, asylums, poor houses—which are spatialized, compartmentalized, hierarchical, and normalizing institutions where persons are studied and learn to internalize the normalizing judgments of experts.

This was Foucault's observation about sexuality—we think sexuality is a thing, something we have or are, but it is not, it is a category of experience constructed in the modern age.[4] Previously, sexuality was understood in terms of behavior: certain acts were condemned, but they did not define "who" someone was. When sexuality became the object of scientific study it succumbed to the logic of scientific discourse—which is directed towards revealing hidden truths and distinguishing "normal" from "deviant," as well as motivated by the reproductive needs of modern industrial society. As a result, "homosexuality" became a pathology and, more importantly, persons became "homosexual." Sexual identity—so crucial to the modern sense of self—is not a *thing* that was studied, it was *created* through the process of study. If one looks to other cultures, such as Arabic, many southeast Asian cultures, and many "indigenous" cultures, the same categories of gender and sexuality do not exist, and there is not the same focus on "coming out" as part of a "true identity." It is not because these cultures are "backwards" and Western cultures are "liberated" or "progressive," it is because sexuality is defined differently in different cultures.

Many thinkers, such as Simone de Beauvoir and Franz Fanon, used similar ideas about discursive power to include gender, race, and colonial discourses, all of which create specific experiences and identities that seem to be "real." This kind of discursive power—now often enforced through discourses of nationalism—is difficult to resist not only because it is difficult to recognize, but because it feels like it comes from inside of us, and we perceive it as normal, natural, given, and obvious. To claim that gender or race is created, is not only counterintuitive, it sounds crazy. This different form of power—bio-power—needs to be addressed and transformed by a different kind of revolutionary action. Most of the chapters in this volume are engaged in tripping up some sort of authoritative discourse about human bodies that has been naturalized through both religious texts and scientific understandings of "nature." In particular the reader should look at the chapters from Catherine Keller and Laurel Schneider, Julia Watts Belser, Teresa Hornsby, Zairong Xiang, Carlos R. Fernández, and Alex Carr Johnson.

Identity Politics: The Emergence of Queer Theory

Queer theory also arises out of and in response to various identity movements, including feminism, gay rights, and transgender/sexual movements.[5] Significant for the development of queer theory is the way that these groups challenged norms of sex, gender, and sexuality, revealing the ways that they are social constructs with historically and culturally specific forms and using historical and other evidence to reveal what had been hidden by the dominant culture—women's histories, gay histories, and so forth. These movements revealed in a concrete way the relevance of the arguments that Derrida and Foucault were making about binaries, exclusion, and the power of discourse. But struggles within these movements also revealed the limits of identity politics—and the modern concept of identity—and opened the door for different approaches to politics and social change.

In identity politics, persons who had been denied rights coalesce around a shared identity which provides the cohesion necessary for political action. Efforts to define and delimit the boundaries of group identity, an effort to empower those who had been marginalized or excluded from public life and discourse, highlighted the ways that identity relies on exclusion and marginalization and reinforces the position of those who have power to name and define that identity. So, for example, women's groups emphasized a similarity between women, often defining women in a way that excluded women who did not conform to standards of white, heteronormativity.

This led to push back, particularly by women of color, those who identified as queer, and, of course, queers of color. Audre Lorde, writing as "a forty-nine-year-old Black lesbian feminist socialist mother of two, including one boy, and a member of an interracial couple" usually found herself part of a group "defined as other, deviant, inferior, or just plain wrong."[6] The problem, is the existence of a "mythical norm"—usually defined as "white, thin, male, young, heterosexual, christian, and financially secure"—in which power resides. Those who stand outside power because of one difference assume that difference is the cause of all oppression, ignoring all the other exclusions and distortions. She goes on to argue that this is the case with the women's movement, in which "white women focus upon their oppression as women and ignore differences of race, sexual preference, class, and age."[7] There is, according to Lorde, a false homogeneity built into the word "sisterhood." The problem is not difference; it is the refusal to recognize difference and to examine the distortions that result from that refusal. A number of other activists and thinkers—notably bell hooks and Gloria Anzaldúa—made similar arguments that by focusing on a single dimension of identity and oppression one masks an enormous amount of difference, which results in the marginalization of others (e.g., women of color in women's groups). This led to a more intersectional analysis which explores the interlocking matrices of identity and oppression. Most significantly, interest shifted from identity politics to a politics of difference. All of the chapters in this volume engage with an intersectional analysis that, while impossible to include all markers of difference, understands the interrelated way in which race, class, ability, sexuality, sex, gender, and even species markers combine in various ways to create a proliferation of different bodies, none of which can be subsumed under clean-cut markers of difference.

The Emergence of Queer Theory

Queer theory emerges in the early 1990s at the intersection of deconstruction, post-structural critical analysis, and political activism, as well as feminist and gay and lesbian studies, in part to take on the complexity of intersecting differences. Although the term "queer" had already been used by activists and intellectuals, the term "queer theory" was first used by Teresa de Lauretis at a 1990 conference at UC Santa Cruz, which resulted in a special issue of *differences: A Journal of Feminist Cultural Studies* (1991) focusing on "Queer Theory." QT inherited certain analytical interests from feminism and lesbian and gay studies—including interest in the ways that gender and sexuality are historically constructed binaries and a desire to explore and challenge the discourses that produced and were produced by those binaries. There are, however, some crucial and interrelated distinctions.

First, queer theory views gender and sexuality as part of a "heterosexual matrix" of interconnected binaries including biological sex (female or male), gender (feminine or masculine), and sexual desire (homosexual or heterosexual), each of which is a social construct. These narrowly constructed categories are intertwined, such that one category "naturally" establishes one's position in another—males are masculine and attracted to females, and so forth. This "logic of heteronormativity," which Eve Sedgwick claims permeates a broad range of discourses, creates assumptions about what is natural and unnatural. Thus, a "masculine" man is "straight-acting," which makes a man who is masculine and has sex with other men incongruous; and muscles are not feminine, so a female athlete might be accused of playing (or looking) "like a man" (which is why female athletes often accentuate feminine gender markers—such as long hair or wearing makeup—in order to appear more feminine and thus avoid accusations of lesbianism). Sedgwick goes on to highlight the power of these constructed gender norms. She notes that of the many ways we might differentiate genital activity–preference for certain acts, sensations, frequency, relations of power and so forth, "precisely one, the gender of the object choice" remains "*the* dimension" by which we understand "sexual orientation."[8]

Unlike gender or sexuality studies, QT does not focus solely on the ways identities such as "woman," "gay," or "lesbian" are formed, or how they become marginalized as other. Queer theory shifts the focus slightly from marginalization to emphasize the ways identities exclude difference and, thus, impact all of us. In other words, rather than examining the ways that constructions of masculinity or heterosexuality create a system of oppression for women, or gay men or lesbians, QT wants to look at and dismantle the way a variety of discourses work to place all of us in one box or another, boxes which fit some of us better than others (and some of us not at all); but which reinforce ideological systems and benefit those with power. Queer theory does not merely criticize the centrality of certain categories, it challenges the very idea of categories. For this reason, QT is a fecund source for challenging boundaries of what counts as "science" and what counts as "religion," not to mention what gets included and excluded from mainstream SRD discussions.

There is, however, tension between gender, sexuality, and other analyses—which are more closely connected to identity politics—and QT, which challenges the modern concept of identity and proposes a more fluid—and to some, threatening—alternative. Many in identity-based politics have, indeed, critiqued QT as dismantling particular political voices that were hard fought for in the public sphere. However, there are also many ways in which QT carries on in the liberative tradition of earlier identity politics. This is perhaps captured best in the work of one of QT's founding scholars: Judith Butler.

Judith Butler: Queering Identities

Several years ago, Lisa was talking with her brother and asked after his roommate and her son, who was around 12 at the time. He said they were fine, although he was worried about Jacob, who (paraphrasing) was too dependent on his mom: "I'm trying to teach Jacob how to be a man." When asked whether girls had the same problem, he explained that girls already know how to be girls—for them it is natural. Simone de Beauvoir would say Lisa's brother was exactly wrong, but Judith Butler might say he got it half right, because knowing "how to be a man" is *not* natural—but neither is knowing "how to be a woman." All of us have to be taught, and we have to keep at it constantly. This is because gender is not a thing, it is, according to Butler, a performative act.[9] Some speech is performative. If I say "I bet you $5," that statement creates a bet; "I now pronounce you husband and wife" calls a marriage into being. Performative acts are acts which—through repetition—create at least the appearance of a thing: a marriage is repeated acts of "being married." The actions required for any given marriage will be dependent on context (as marriage has certainly changed over time), and the individuals involved. But, whatever the acts, if one or both partners stops performing those acts—if they stop acting like they are married—the marriage ceases to exist (we might say it was "a marriage in name only"). Gender is like marriage in the sense that it begins with a speech act "It's a girl!," but is maintained by repeated acts of "being a girl." It is by acting "like" a woman that I become, and remain, a woman. This repetition gives the appearance of a gendered self—just like the repeated acts of a marriage make the marriage appear—but there is no gender apart from the act. Gender, like marriage, is fragile, and we can blow it all with a failed performance, which is why infidelity can threaten a marriage, and why many female athletes are so vigilant at performing femininity.

The fragility of gender—and the necessity of policing the boundaries of performance—mean that we all experience "gender trouble," the name of

Butler's seminal work in this area, exerting an enormous amount of time and energy—both personal and social—maintaining a system that is anything but natural and fixed. As an alternative, Butler wants to deconstruct traditional categories of identity by asking how and why they were created, what has been hidden to make them appear, and how they come to seem "real, natural, and universal."[10] There are many ways to do this, although her most well known example might be through alternative performances of gender, such as "drag," where the imitation of gender reveals its performative nature and, through the subversion of such things as parody, which exposes the discontinuities between sex, gender, and desire.[11] Since Butler's seminal work in this area, there have, of course, been many more developments in the field (some of which will be discussed in this book), but all of these developments hinge to some extent on the "de-naturalizing" of categories such as gender, sex, and sexuality.

QUEER SCIENCE

Queer theory has also influenced, and been influenced by, the natural, physical, and biological sciences. The natural world exhibits an enormous degree of diversity in biological sex (many species switch between "male" and "female"), gender traits, sexual behavior, and modes of reproduction,[12] all of which undermine the notion that sex and gender are binary and that sexuality is dominated by heterosexual forms. This diversity was obscured by the discursive frameworks of the natural sciences, which were shaped by the interlocking binaries of sex, gender and attraction—male and female; masculine and feminine; heterosexual and homosexual. Scientists expected to see heterosexual behavior, which was understood as "natural," so they described any contact between a male and female as sexual. Because they did not expect to see homosexual behavior, virtually no contact between animals deemed to be of the same sex was described as sexual or, if it was, it was described as an aberration, a deviance from what "should" be happening. Ideas about what is "natural" thus permeate descriptions of the reproductive behavior of many species, such as a description of the New Mexico Whiptail, an all-female species of lizard—often labeled the "lesbian lizard"—which reproduces by parthenogenesis (eggs develop into embryos without being fertilized). Whiptail "virgins" are described as lacking the stimulation which a male should provide, so another female "forsakes her true gender," "apes being a male," and "goes through the motions" of copulating.[13] This illustrate the relationship between scientific discourses and what is deemed natural and deviant—note how heteronormative ideas are intertwined with the observations—the "lesbian" lizard lacks a man, forsakes its "true gender," and somehow mimics "real" (hetero)sexual behaviors. We should also note the incongruity of describing the sexual behavior of virgins. Alex Carr Johnson's chapter, in particular in this volume, addresses some of these issues and how some in conservation biology reinforce heteronormative understandings of nature with the "science" that they do.

This is not to say that queer science is merely an object for queer analysis. Science has certainly contributed to the growing perception of how queer the world is, and challenged our certainties about the language that we use to describe it—indeed, this is how science is supposed to work! These discoveries push queer theory to go further, because if Judith Butler suggested that the whole system of identity is unstable because gender is unstable, the natural sciences demonstrate that reproduction and biological sex are also unstable and constructed. Anne Fausto-Sterling, one of the leading thinkers in queer theory and science, describes the phallo-metrics of birth: when you were born, if your genitals were less than 3/8 of an inch long, you were deemed to be a girl; if it was longer than an inch, it was a penis and you were a boy. Those whose genitals are in between are "deformed," assigned a sex, and surgery is performed to "fix the problem." There was no other option because doctors "never questioned the fundamental assumptions that there are only two sexes"[14]—they cannot see alternatives because those alternatives do not exist in their framework. Thus, a significant number of people—between 1 in 1500 to 1 in 2000—are "fixed" to fit the male/female binary.[15]

Evolutionary theory itself can also be read queerly. Joan Roughgarden, for instance, has questioned the heterosexism read back into the natural world in the form of survival of the fittest, and argued that far from being all about competition and finding the right "mate" to propagate one's genes, evolution is also a story about incredible and unbelievable cross-species and multi-organism cooperation. Most species exist because multiple other organisms pave the way and/or fall into place to create a "niche" for a given organism to thrive. Hence, reading heterosexual reproduction and "battle" metaphors into evolution has obscured these more cooperative tendencies within the history of evolution.[16]

The story of evolution is also the story in which humans are made a part of the rest of the natural world—putting humans in continuity with other life on the planet. This very move "queers" the place of humans because we are no longer "above" the rest of the natural world, which is a metaphor that is still dominant in many philosophical and religious ideas. Furthermore, it can be argued that the real story of evolution is that there are no real distinct "species" in nature; life evolves from one form into another and our designation as this or that species is merely nomenclature and conceptually useful for humans to understand the world in a given place and time. This blurring of the boundaries between species and organisms in general, between humans and animals, between plants and animals, and between biotic and abiotic life is surely a useful tool for "queering" our understandings of the natural world. Discussions in neuroscience about the neuroplasticity of individuals—in which our brains are literally changed by each and every interaction with "others" in the world—and in microbiology—which tells us that only 1 in 10 cells of an individual are human cells and understands each individual as an ecosystem made up of a microbiome—also do much to "queer" our understandings of nature, bodies, and individuals.[17]

The same can be said of the science of physics. At least since Einstein's relativity theory, scholars have been pondering what it means that matter and energy are the same sort of thing: matter is really just a "collection" of energy in a space-time fold. Or, put differently, that the reality of life is that everything is made up of energy-events. Process thought built up a model based on the science of the quantum world, and one of the most interesting things about this model is that it suggests all life is constantly in flux and that life is in a very "real" way interrelated.[18] Ideas about string theory and the multiverse emerging out of quantum physics and cosmology are also quite "queer" in that they really challenge our understanding of a stable reality and single universe. In this way, we might say that these sciences "queer" the everyday experiences of reality that we have.

More recently, Karen Barad has directly brought queer theory into dialogue with quantum reality and suggests that something like Judith Butler's notion of "performativity" is at work "all the way down" and "all the way out." Many of the authors in this volume bring up Karen Barad's groundbreaking work in her book *Meeting the Universe Halfway*. There she argues, we ought to think of the systems and patterns of nature as also "performative" to some degree.[19] While not all of the authors in this volume would agree, several would. In other words, she is arguing that even at the quantum level there is some type of agential reality (what she terms "agential realism"), and that even at that level there are possibilities for reality to become in different ways. Laws, then, are not so much determined and written in stone as they are deep habits of universal becoming. These habits for becoming, and the agential realism she argues for permeates all of reality and extends to the expanding edges of the universe: it challenges the teleology within nature and within the whole cosmos. Things do not have to be "as they are" but every "level" of life is pregnant with possibilities for becoming.[20] Instead of thinking of "nature," "reality," and the "universe" as unfolding in any particular or determined way, all of reality is much more like a starling murmuration, moving this way or that in definite patterns, but with no ultimate telos. This is a far cry from the reality that most of us take for granted on a daily basis. In a way, it is a queering of reality at a very fundamental level to suggest that there is no "just the way things are" even at the level of "nature," but rather, nature is also always in a process of becoming and changing. From where we stand, it may seem stable, but such stability should not be confused with reality as it is and always will be.

In these ways, and many more, "science" itself provides resources for queering our understandings of the natural world of which we are a part. Of course, the practice of science is often engaged in telling us "the way things are," but science at its best recognizes, even within its methods, that science is never giving us a mere description of data about the world, but rather is engaged in trying to find the best theory that takes account of the most data in a dynamic, changing world. How we see the world and the theories we use to examine the world, help shape the worlds that we co-create. Such an understanding of the relationship between data and theory, as Barad notes, is much closer to Niels Bohr's understanding of the world as indeterminate, than it is to Heisenberg's understanding that eventually our knowledge will discover the true reality of the world. In other words, science is much queerer than it might appear on the surface. The same is also the case with religion and religious traditions.

QUEER RELIGION

It may seem odd, at first, to insist that religion is much queerer than meets the eye. After all, what has been more normative for gender and sex roles than religious ideas and traditions? According to one interpretation of the monotheistic creation stories, God created male and female for procreative purposes. In Vedic cultures and in Chinese religions as well, gender dimorphism and heteronormativity have been assumed the norm.[21] However, queer theory argues that such readings are just that: readings from a particular, historical social location, and one that may be more influenced by Victorian ideals of gender, sexuality, and family than the realities out of which religious traditions emerged.[22]

Luckily for the sake of this volume, and for many peoples around the planet, the interpretations of religious traditions are just as polydox and multiple (if not more) as are readings of nature. In particular, the chapters from Catherine Keller and Laurel Schneider, Julia Watts Belser, Zairong Xiang, and Fern Feldman deal with interpretations of religious texts and theological ideas, while Teresa Hornsby deals with how apocalyptic and dualistic understandings of reality from theological traditions permeate even pop culture.

Reading religion as resistance to established norms is "an interesting starting point to study religious practices, identities and ethics."[23] In this brief space, we offer some queerly resistant readings of the histories of religious traditions, of theological and philosophical issues within religious traditions, of religious texts, and of religious ethics. These are cursory introductory remarks as each one of these foci in queer theory and religious studies now has multiple volumes, theses, and articles written about it. But for those who are unfamiliar with the study of queer theory and religion, they may offer some guidance to the text that follows.

First, it should be noted, that queer identities litter the landscape of religious traditions. This was one of the primary points of John Boswell's work in *Same Sex Unions in Premodern Europe*, for instance, which shows same sex loving relationships throughout the history of the Christian west.[24] It may seem counterintuitive to "look for" queer identities in history when queer theory argues in part that identity structures and the nature of sex and sexuality change over time, but nevertheless, such historical examinations of other than hetero-relations help to loosen gender and sex identities from the confines of the present. There is also much evidence of same sex relations within Christian, Buddhist, and other monastic communities, as is evidenced by love letters, poetry, and historical documents condemning such behavior.[25] Furthermore, in South, Southeast Asian, and many Indigenous traditions, "third" and "fourth" genders (what we might now call trans* in some way) have often played important religious roles throughout history.[26]

One also finds many queer identities within religious texts themselves. For instance, one common reading of the Garden narrative in the Genesis story has assumed that God created Adam and Eve for procreative purposes (a form of heternormative gender dimorphism). However, upon closer scrutiny, Adam was created from the adamah (mud) from God, and was created as a gender-less "mud creature." Eve (life) was then created as a helpmeet (an equal companion) to Adam.[27] The *Kama Sutra* also relays stories of same sex desires (Part II, Chapter 9). The Buddha speaks of four genders (not two): male, female, "effeminate males," and hermaphrodites (*Vinyana*) while Vedic texts speak of the Tritiya-Prakriti (third genders) who were (again) often important in religious ceremonies. In other instances Islamic and Christian and Buddhist scholars, monks, and poets would write of same-sex relations (whether in adoration or condemnation), which belies the fact that these relationships existed and were a part of the cultures that mutually influenced these religious traditions and in which they were housed.[28] The point here is that when one looks closely at religious texts from a non-heteronormative lens, one begins to understand that a variety of identities populate the pages of these religious/historical texts in both explicit ways and in understanding what is left out of the texts: for example, an admonition to perform sexual acts in a certain way obviously means that sex was happening in many different ways. Having said that, it is important not to anachronistically read contemporary understandings of sex, gender, and sexuality back on to texts. As many have pointed out, patriarchal Rome in which many New Testament texts were written, for instance, understood one proper gender (the patriarchal male) and all others were defects of that ideal.[29] This is much different from the gender dimorphic, heteronormative context of Victorian family values in the Euro-Western world today.

Yet another way in which one can understand religions queerly is through their theological and philosophical understandings of the relationships between humans, animals, god(s), life, and death. As Marcella Althaus-Reid reminds us, every version of theological understandings of the world is just that, a version: per/version.[30] We ought to understand better how our versions of theology write our sexes, genders, and sexualities: how they help us to relate to other human beings. Rather than searching for an orthodox understanding of these relations, we may do better by understanding the polydoxy of every religious tradition: at a given moment in time, there has never been a single answer to theological questions nor for how we ought to relate to one another (sexually or otherwise).[31] Furthermore, we might look to our theological and philosophical understandings of reality as themselves, quite odd. The trinity is three persons in one (perhaps no better metaphor for relational understandings of identity); Christ is both human and divine (incarnation blurs the boundaries between gods/humans); trickster figures in many indigenous traditions blur the boundaries between living/dead, male/female, and human/animal; and the idea of dependent co-arising in Buddhism suggests, along the lines of Deleuze and Guattari two thousand years later, that the virtual is the real: the interrelated flux of things is what is really real, while individual bodies and things are abstractions from this evolving flux.[32] In all of these ways and more, concepts about ultimate reality and/or god/s in the world's religious traditions paint a much queerer picture of the world than the taken-for-granted everyday lives we often live.

Finally, religious ethical systems and admonitions are also a source for thinking queerly about relationships within the world. Jesus' admonition to his disciples to leave their families behind and begin a new family that would include male/female, Jew/Gentile is a queering of the ethical boundaries of concern at that time. The radical ahimsa found at the heart of Jainism which extends to all entities in the world, expands the boundary of moral concern past that of the human realm to include other animals and plants. Tree ordaining monks, saints that speak to animals, incarnational ethics that suggest all of the elements of the world (and even universe) are part of the divine and thus deserve our ethical and moral attention; these boundary queering ethics and more are all found within the world's religious traditions. In a sense, as with understandings of the natural world through the sciences, all we need to do to grasp the queerness of the world is to pay more attention to what the world's religious traditions have to say to us. Many of the chapters in this volume will draw from these and other queer traditions within the world's religions to help bolster and illustrate their arguments. We leave the reader to those chapters to read more deeply about QT and religion, so that we can move on to the final section of this introduction: the Religion and Science Discourse.

THE RELIGION AND SCIENCE DISCOURSE

Background

As the preceding sections have argued both religions and sciences are plural, living traditions made up by and of human beings. In other words, there is no abstract "religion" or "science" out there that have come together to have dialogue or arguments over tea. Rather, these traditions and institutions find themselves always already together in the minds, hearts, and bodies of individual human beings. In a sense, telling the story of "religion and science" is telling the story about how everyone navigates the different types of knowledge these methods and institutions provide for constructing some sort of meaningful understanding of the world. In other words, we would argue that there are very few people (if any) who use only either religious or scientific methods for knowing the world—it is not even clear what that would look like since "religious" and "scientific" traditions are historically intertwined. Most of the chapters in this volume, then, "think" these two things together, although the chapters from Clayton and Florez, Stenmark, and Bauman deal more closely with the field that is called "science and religion."

The science and religion discourse emerges in the particular context of Western thought and, more particularly, from the somewhat recent perception that there is something called "science" that is distinct from "religion." What we know as modern Western science begins largely with the natural philosophy of the Ancient Greeks—Aristotelian or Ptolemaic cosmology, with the earth at the center and lowest point, and the planets and sun orbiting the earth on concentric, wider, and higher circles moving outward and upward from the earth and culminating in the realm of the fixed stars. Christian theology became aligned with Greek cosmology, and a good example of this is the concept of the Trinity and the nature of Christ. For the ancient Greeks, movement was corruption. God or the Real, could not be corrupted or unchanged and thus could not be present on the earth. How could Christians make sense of their God coming to earth and being embodied? By the time of Augustine, the Trinity was thoroughly entrenched in this worldview: Christ was a combination of two forms and one body, or two substances and one form; God the Father was the transcendent creator and sustainer (like the unmoved mover); and the Holy Spirit mediated between God and the world (like the cosmic world soul).[33] Thus, natural philosophy, or "proto-science" and theology were intertwined from the beginning.

The "Dark Ages" was, of course, the Golden Age of Islam during which the Islamic Abbasid Empire became the seat of developments in natural philosophy and science. Developments in early medicine, optics, chemistry (drawing from Chinese alchemy), algebra (from Vedic and Chinese cultures), among others took place all in Arabic during the period from about 600–1200 CE. Some of the pillars of Islam even fueled some scientific advances. For instance, the need to pray to Mecca led to advances in technologies that helped one determine the direction of Mecca; the need for everyone to read the Qur'an helped spread literacy, which also meant more people had access to natural philosophy; and the Hajj meant that peoples from all over the empire would come together in pilgrimage and share ideas from various parts of the world. As the Renaissance in Europe began, it was fueled by these Arabic translations of Greek texts and the advancements that had been made during the Golden Age of Islam. In this sense too, modern Western science is an effort from multiple cultures and traditions around the world.

What happened with Copernicus—and later, Galileo—during what is often referred to as the "scientific revolution" was a challenge to two thousand-ish years of understanding the cosmos, and one thousand-ish years of understanding Christian theology in the West. These ideas were not just a challenge to theology, but also a challenge to our understanding of the natural world. Our senses, it turns out, can deceive us. This challenged the existing understanding of the relationship between the Two Books—Scripture and the Book of Nature. These two books were not in conflict, and Augustine (354–430) suggested that the book of nature was more accessible because even those who were illiterate could read it. But it began to seem like nature might not be all that easy to read. We needed experts there too because nature was deceptive, and her secrets needed to be unlocked through active intervention, not passive contemplation. This led to a vast shift in perception, a "death of nature" in which nature was stripped of religious significance, and in which a contemplative approach towards knowledge of the natural world was replaced with the utilitarian goal of achieving mastery over nature.[34]

Other important shifts included the perception of the world as a machine—not an organism—and a growing sense that there were universal laws that ordered the movement of the world and its parts. And, finally, an extension of natural laws into the human world, which included applying scientific methods to ethical and political questions—for example, Thomas Hobbes's *Leviathan* (1651) and David Hume's "A Treatise of Human Nature" (1739–40)—but also led to the emergence of the social sciences (which so interested Foucault). The social sciences simultaneously rejected the idea of differences among humans—"human nature" is singular—while nurturing a schizophrenic obsession with defining cultural and racial differences in order to identify and explain Western distinction and superiority.

During the 19th century, travel helped strengthen the idea that science gave us some sort of objective understanding of the world, while religion provided us with subjective or parochial understandings of the world. As people such as Goethe, Alexander von Humboldt, Ernst Haeckel, and Charles Darwin traveled the world, they began to see "patterns" in nature that science could make sense of, while at the same time they saw local customs as varied and subjective, that is, as not universal.[35] Darwin's travels led to his theory of natural selection, and filling in this narrative not only relied on travel, it relied on the cataloging of species made possible by indigenous peoples. We would not have the rich understanding of botany, medicine, species diversity, and biodiversity were it not for the indigenous knowledges that contributed to these sciences. Yet, because this knowledge was seen as part of a non-scientific worldview, it was treated as "discovered" by Europeans and thus reinforced the idea that Europeans had some special relationship and method to understand nature, while all others would have to be brought into this "enlightened" fold.

Finally, just as Darwin and Copernicus before him helped transform religion and science, so too did Einstein in the 20th century with his revolution in quantum physics. But, again, not without being in dialogue with the world's religion. By the time of the 19th century, physicists such as Bohr, Heisenberg, and Einstein were beginning to understand that Newtonian mechanics did not tell the whole story. In part their readings of the world were influenced by the influx of "eastern" understandings of the world during the 17th to 19th centuries. In other words, whereas the metaphysics of the monotheistic West had been shaped by the separation of things, of matter and energy, the metaphysics of the East—whether Taoist, Confucian, Hindu, or Buddhist—were marked by an understanding that things were inextricably related; that matter and energy were at heart one. These metaphors helped shape the wheelhouse of possibilities for scientists examining the physical world and cannot be seen as isolated from the shifts that happen in quantum physics of the 20th century.[36]

Regardless of how we may tend to understand "religion" and "science" as separate entities, particularly in the Christian West, the history of science and religion show that science continuously defines itself over and against—that is, in relationship to—the religious tradition(s) it emerges from. Thus, throughout this entire period, but particularly in the 19th and 20th centuries, there was an ongoing attempt to describe "science" as a particular kind of knowledge, distinct from other knowledge-seeking endeavors. Science was not superstition or magic and was distinct from indigenous traditions (Malinowski, 1925), the Humanities (Thomas Henry Huxley,1880; C.P. Snow 1959), or "pseudoscience" (Karl Popper, 1959). And, it was certainly not religion. This process of separation and the assumed supremacy of either religion or science helped contribute to the idea of a "warfare" between religion and science. Perhaps the most important articulation of this position appeared in the early 1870s with John William Draper's *History of the Conflict Between Religion and Science*, followed by Andrew White's *A History of the Warfare of Science and Theology in Christendom.*[37] In 1927 the Science League of America, described science and religion as "two opposing cultures," one of which must dominate the other: "On the one side we see arrayed the forces of progress and enlightenment, on the other the forces of reaction, the apostles of traditionalism. There can be no compromise between these diametrically opposed armies."[38] What is significant to note is that most of this rhetoric comes from those defining science, an effort which inevitably involves defining what science is not.

"Science and Religion" as an Academic Discourse

This is the context in which the current religion and science discourse emerges in the 1960s as scholars began reexamining the relationship—largely influenced by work in philosophy of science, such as Kuhn's *Structure of Scientific Revolutions*. This really started with Ian Barbour's *Issues in Science and Religion* (1966) and *Myths, Models and Paradigms* (1974) which built on Kuhn's work and focused largely on methodological issues: viz. that both are theory/model dependent, rely on metaphors and models, value coherence, comprehensiveness, and fruitfulness. Other scholars began exploring the similarities between religion and science in their methods, institutional communities and traditions of knowledge production that depend upon authoritative "experts." Furthermore, both science and religion have moments of change—called reformations and prophetic voices in religion, and new discoveries and paradigm shifts in science—and moments of dogmatic refusal to accept new information.

The focus began to shift in the 1980s to include historical reexaminations of the relationship between religion and science, challenging the warfare model by demonstrating the ways that what we now call religion and science have always been intertwined. There were also attempts to address—and resolve—perceived differences between religion and science, focusing on truth claims, particularly the conflict between "creation" and "evolution." By the end of the 20th century, scholars had begun to explore different models for understanding the relationship between religion and science. Here again, Barbour was pivitol. *When Science Meets Religion* (2000) describes four ways that science and religion can be related to one another: a) a conflict model in which only one is correct; b) an independence or two world's model in which science and religion represent two distinct practices (Stephen Jay Gould referred to this as "Non-Overlapping Magisteria"); c) a dialogue model in which science and religion are separate fields with some commonalities and conflicts, that can come together to address specific issues; and d) an integrative model in which science and religion only seem to be saying different things, but they will eventually converge into a common understanding of reality.[39]

Most recently there has been a shift to focusing on issues and ethics, such as the environment, bioethics, and industrial agriculture and food. Fifty years ago, Lynn White Jr. argued that the environmental crisis is not just a problem of technology, but a religious problem as well.[40] We can say the same today of climate change. Climate change is not only the result of fossil-fueled technologies that enable a huge increase in the speed of communication, transportation, and production, it is also a crisis about who we are in relation to one another and to the more than human world. It forces us to question the material ways in which our technologies and behaviors impact the rest of the natural world, and our place as human beings within a planetary community. How we value other life forms, future human generations, and the ecosystems of the planet will inform whether and how we address climate change.

The same can be said for end of life issues that have fallen historically under the rubric of bioethics. Should physician-assisted suicide be legal? If so, for whom? How many resources should we spend keeping one person alive at the end of his/her life? Should people die in hospitals? Or, might death (and birth for that matter) be better done at home? At what point does life begin and up to when should a woman have the right to choose? What happens to children born intersexed? Should their parents be allowed to choose their child's sex at birth, allow them to remain intersexed, or should we allow for children to make their own decisions about this once they reach physical maturity? These are all questions that neither science nor religion alone can answer, because they involve various issues of meaning and value, as well as scientific understandings of what life and death, and male and female might mean. Leaving these questions up to science and medicine (and thus insurance companies) alone means the most economically expedient and culturally accepted procedures will be favored. At the same time, leaving them up to religion alone might ignore the unique, embodied experience of a given individual and favor some type of general principles or values. These are questions involving religion *and* science (and other discourses as well).

Finally, we can examine questions of food production. From the ethics and technologies of eating other animals, to the ethics and technologies of farming practices—GMOs, monocultures, pesticides, and fertilizers—scientific and religious/ethical decisions go into every meal that finds its way to our tables. How are farm-workers treated who raise and pick the crops that you eat? How far does your food travel and what is its overall ecological footprint?[41] Should the water and air necessary for food production—and thus for all life—be held in common? What about holding the land in common, as it was up until the enclosure of the commons in the 19th century? How can technologies help and hinder sustainable and fair-trade practices in the foods that we produce? All of these questions go into every bite, yet we rarely take the time to discuss where our food even comes from.

Historically, the academic discourse of science and religion has been dominated by theological, Christian, and Western concerns, which has to do, in part, with the reality that it is in Christian, Western cultures that science and religion become conceptually separated and in which they must come back together. Many Eastern traditions and indigenous traditions (and Judaism and Islam to differing degrees) do not see religion as something that is separated out from the rest of public life, including science, economics, and politics. But, this focus also has to do with economics and power: Christian nations have for the past century and a half been spreading (imposing) "Modern," and "Industrial" ideas upon the rest of the world (see especially Stenmark's chapter on this point). The largest funders of projects within science and religion, such as the John Templeton Foundation, also come from these Western, Christian-theological traditions and tend also to fund the types of concerns that emerge out of Western contexts. This is a problem within the field that is beginning to be addressed, as we all come to terms with the realities of globalization and climate change.

QUEERER STILL: THE FUTURE OF SRD

As the above suggests, religion, science, and the SRD are already queer and they always have been: challenging the boundaries between religion and science, exploring and reinterpreting history. The boundary between "science" and "religion" is now and always has been shifting, blurry, and porous, and it is the location in which the SRD makes its home. There is more to be done of course. It is our hope that this volume, which is the first in a new series by Lexington Books entitled "Religion and Science as a Critical Discourse," will begin to broaden what the SRD might be to include ideas and concerns that have not heartily been taken up by the SRD to date.

Too often we are drawn to preserve the status quo, we are not inclusive enough, and we have—and offer—too narrow a view of what is included in the discourse. Thus, while the SRD is already queer, it needs to be queerer still. We think that the methods of science and religion are up to this task, but while we are in debt to earlier models, none of them quite capture the "always-already togetherness" of science and religion for which we have been arguing.[42] Lisa has, for instance, argued for a model of "disputational friendship" in which we understand that both discourses influence our daily lives and the political sphere, but also in which we might be able to use the polis to explore the possible implications of multiple different points of views concerning issues of science and religion, in an attempt to critically examine how we each construct meaningful worlds using both scientific and religious information.[43] Our bodies are all differently located and differently formed in histories of nature-cultures, and they experience scientific and religious worlds differently, and these differences matter.[44] They must be navigated in dialogues in which we seek to understand the other before jumping to criticism.

We are not arguing for an uncritical appropriation of queer theory, and the chapters in this volume suggest some of its limitations. For instance, Emilie Towne's chapter draws from the often neglected field of womanist studies to address some of the limitations of QT and SRD. Julia Watts Belser does the same with critical race theory and feminist disability studies, and Lisa Stenmark does the same with post-colonial theories. Engagements with discourses that are not often taken up by QT and/or SRD help to expand the conversation and cast a wider analytical lens on the problems at hand.

One significant criticism of QT is that it has been dominated by white, middle-class concerns: "Much of queer theory nowadays sounds like a metanarrative about the domestic affairs of white homosexuals. Surely, queer studies promises more than a history of gay men, a sociology of gay male sex clubs, an anthropology of gay male tourism, a survey of gay male aesthetics."[45] This is not to say that queer theory has only been the purview of white interests or white academics—José Muñoz' *Disidentifications: Queers of Color and the Performance of Politics* is considered one of the "founding" documents of QT, and even before queer theory was a thing, women of color, such as Audré Lorde and Gloria Anzaldúa were using their experiences to challenge the boundaries of any particular lens: "I am all races because there is the queer of me in all races," writes Anzaldúa, a position that challenges all givens, not merely a creature of light and dark, but one "who questions the definitions of light and dark and gives them new meanings."[46] Thus, many of the chapters in this volume draw from postcolonial, critical race, womanist, disability, and other "critical" discourses to broaden the conversation beyond the historical scope of the Western SRD.

We think it is important to push back against any understanding of queer theory that is too focused on the kinds of parochial concerns that Eng lists above. One way to do this is to keep in mind other critical discourses—including queer voices of color, postcolonial and critical race theories, and other intersectional analyses—not because queer theory is wrong, or these other frameworks are better, but because identity, systems of power, oppression, and injustice are all far too complicated to approach from just one perspective. But, it is equally important that these other concerns and trajectories are not merely subsumed into queer theory, or that they become utilitarian add-ons. These various critical theories share an intellectual debt to such thinkers as Jacques Derrida and Michel Foucault, to Franz Fanon and Edward Said, along with feminist analyses of the relationship between power and construction of social reality, and radical social movements.

It might be helpful in this endeavor to distinguish between narrow and broad understandings of queer theory. Narrowly, queer is "an umbrella term for a coalition of culturally marginalized sexual identifications"[47] and queer theory focuses on the analysis of the relationship between male/female, masculine/feminine, and heterosexual/homosexual, the way that these are intertwined. But, queer theory in its broader sense, extends these critiques and thus questions all norms, and explores and challenges all discourses of power. David Halperin puts it this way:

> Queer is by definition *whatever* is at odds with the normal, the legitimate, the dominant. *There is nothing in particular to which it necessarily refers*. It is an identity without an essence. "Queer," then, demarcates not a positivity but a positionality *vis a vis* the normative. [I]t describes a horizon of possibility whose precise extent and heterogeneous scope cannot in principle be delimited in advance.[48]

Keeping in mind multiple trajectories is especially important for challenging the myopia that can come from single points of interest—for this reason, those of us who identify as queer in the narrow sense, will be served well by entering into conversations with positions less likely to identify with historical centers of power such as race postcolonial theories. In this text, the authors draw from both the "narrow" and "broader" sense of QT. Though we do think the broader sense is a fruitful, useful, and meaningful direction for QT, the "narrow" sense of the term should never be ignored or forgotten as it signifies the histories from which this type of theory emerges and it also keeps the bodies and identities of "queer" peoples in mind, which are still disproportionately negatively affected throughout the world by heteronormative understandings of gender, sex, and sexuality.

We think that this volume will contribute to the ongoing queering of the religion and science discourses. This includes critical contributions, that challenge the boundaries and authorities of disciplines and fields, including QT and the SRD, and that promote alternative metaphors for thinking about what counts as knowledge in both science and religious discourses, and in SRD and QT as well. Finally, we think that one of the unique contributions this text embodies is an overall shift in thinking about epistemology from that of concerns with ontology and metaphysics, and toward that of being concerned with hermeneutics, interpretation, justice, and ethics. In other words, if knowledge and power are related, then we ought to be asking how certain understandings of the world and certain claims of knowledge within the world privilege certain bodies over others. Furthermore, we ought to begin to imagine together, with multiple others both human and non-, what types of worlds we want to live towards. These are questions about justice and ethics more than anything else, and both science and religion will and must contribute to their answers.

Chapter by Chapter Analyses

One of the dominant themes of queer theory—perhaps its defining characteristic—is that any organizational scheme is arbitrary, and the organization of the chapters in this book are testament to this: there are many themes that we could have emphasized, and most of the chapters could fit into multiple categories. Nevertheless, we have divided the chapters into three major categories: Methods and Histories (1–4), Embodied Texts and Theologies (5–10), and Practical Engagements (11 and 12).

In Chapter 1, Clayton and Florez argue that both science and religion provide examples of gender fluidity. They argue that gender fluidity does not merely destabilize fixed conceptions of gender and sexual identity, but rather it also provides insights into fluid identities in religious belief and practice and suggests fluid interpretations of what science can mean for other areas of human existence and practice. Townes extends this theme of destabilization, using the interplay of queer theory and womanist methodology—Quar Theory— as an interlocutor that blurs the boundary not just between religion and science, but also between the academy and the wider community, religion and the secular, and theory and practice. Quar Theory moves us out of our narrow academic boxes and suggests what queer theory has to offer our broader culture(s). In Chapter 3, Stenmark observes that while Europe experienced three revolutions in the 15th and 16th centuries—the Protestant Reformation, the Scientific Revolution, and Global Colonization—the SRD focuses only on the first two, while largely ignoring the third. This reflects a tendency within the SRD to view "religion" and "science" as discrete practices, isolated from broader cultural and historical developments. This practice also exaggerates the conflicts between Western religion and Western science, and obscures their complicity in colonialism. Finally, in Chapter 4 of the "methods and histories" section, Bauman argues against epistemic and ethical dogmatism in both the sciences and religions, positing a queering of authority so that a more dialogical and fluid approach to religious and scientific discussions might emerge. In this framework, knowledge is not built on dogmatic claims of certainty, but emerges from failure, uncertainty, and unknowing as we attempt to cobble together common epistemological grounds from multiple perspectives within the planetary community.

Chapter 5 begins the Embodied Texts and Theologies theme of the book. In it, Wayne White explores the rich conceptual space opened by James Baldwin's use of the bastard metaphor and his ensuing concept of love. "Bastard" is a trope that marks the emergence of an African-American religious naturalism that resists normative (and thus impoverished) views of our shared humanity. This African-American religious naturalism invites readers to reconsider who and what we are: value-laden natural processes that become human in specific orientations. Hornsby, in Chapter 6, focuses on the figure of the Slenderman, an Internet character who exists in the shadows, stalks children, and appears in dreams before victims disappear. Her exegesis: 1) identifies the threat of the old apocalypse (hierarchical dualism) that brings about genocide; 2) explores the ways the trans body's apocalyptic representations simultaneously threaten and redeem; and 3) explores the means to thrive after the demise of the constructed, two-gendered, hierarchy. In Chapter 7, Keller and Schneider offer an ecological and erotic relational ontology. Following Karen Barad, they argue that "nature's queer performativity" is fundamental to an adequate relation to ecological/planetary forces—to "spacetime mattering," or what they call "promiscuous incarnation." This queerness of desire may, along with Bauman's "polyamoury of place," update Whitehead's "Eros of the universe" as the basis for a theological "ecosmopolitics." In Chapter 8, Watts Belser brings queer theory into conversation with critical race theory, feminist disability studies, and animal studies to scrutinize the Babylonian Talmud's instructions to recite a blessing praising God for creating diverse creatures when confronted with a strange or unusual individual. She analyzes the afterlife of this rabbinic blessing in later Jewish literatures and probes the dangers present in a queer move to destabilize the "naturalness" of difference. Chapter 9 moves to the creation story in Genesis 2. Xiang analyzes the creation(s) of "man" through an agglomeration of a literary close-reading of Traditional Chinese Medicine (TCM) and "quantum queer theory." Examining the sexually entangled body of *'adam* with the help of the model of "sexless gendered body" theory that he develops, he suggests that the biblical text, far from proposing a sexist theology of male dominance, actually retains an ambiguous space of constant queering. Finally, rounding out this section, Feldman weaves together Jewish mystical texts, Karen Barad's theory of agential realism, and Judith Butler's theory of gender performativity, in order to break free from identity binaries. This reading of Hebrew Scriptures and Jewish mystical texts provides a queer theological understanding of the Jewish mystical view that there is nothing but the One, in all its many manifestations.

In the final section of the book, we conclude with a few practical applications as examples of how queering the SRD might materialize in the world. Carr Johnson, in Chapter 11, notes that while the field of conservation biology has developed strategies for accomplishing seemingly insurmountable tasks, these strategies often, wittingly or not, reinforce hegemonic conceptions of gender, reproduction, biological fitness, and conservation ethic. He argues that challenging conservation's heteronormative biases might increase the field's capacity to accurately assess and understand biodiversity and the complexity of preserving it, as well as inspire a more-inclusive conservation ethic. In Chapter 12, Fernández argues for a queering of information science. The chapter scrutinizes antiquated cataloging rules based on deep-rooted societal views, which make research difficult. For instance, some libraries still catalog books on homosexuality and transsexualism under medical and psychological disorders and deviancies. The chapter explores the possible benefits that "queering" LGBTQ subject headings will have on users, libraries, and academia. Finally, rounding out the volume is an Afterword by Morton, who argues that we need to think with Ponty about the embodiment of knowledge and how science and spirituality can help to hold places open for new insights to emerge.

WORKS CITED

Althaus-Reid, Marcella. *Indecent Theology: Theological Perversions in Sex, Gender and Politics*. New York: Routledge, 2000.

Anzaldúa, Gloria. *Borderlands/La Frontera: The New Mestiza*. San Francisco: Spinsters/Aunt Lute, 1987.

Bagemihl, Bruce. *Biological Exuberance: Animal Homosexuality and Natural Diversity*. New York: St. Martin's Press, 1999.

Barad, Karen. *Meeting the Universe Halfway: Quantum Physics and the Entanglement of Matter and Meeting*. Durham, NC: Duke University Press, 2007.

Barbour, Ian G. *When Science Meets Religion: Enemies, Strangers, or Partners?* New York: Harper Collins, 2000.

Bauman, Whitney. *Religion and Ecology: Developing a Planetary Ethic*. New York: Columbia University Press, 2014.

———. "Religion, Science, and Nature: Shifts in Meaning on a Changing Planet." *Zygon: Journal of Religion and Science* 46, no. 4 (December 2011): 777–792.

Biosvert, Daniel and Jay Emerson Johnson, eds. *Queer Religion, 2 Vols*. Santa Barbara, CA: ABC-CLIO, 2012.

Boswell, John. *Same-Sex Unions in Premodern Europe*. New York: Vintage, 1994.

Butler, Judith. *Gender Trouble: Feminism and the Subversion of Identity*. New York: Routledge, 1990.

Deleuze, Gilles and Felix Guattari. *A Thousand Plateaus: Capitalism and Schizophrenia*. Minneapolis, MN: University of MN Press, 1987.

Derrida, Jacques. "Racisms Last Word." *Critical Inquiry* 12 (Autumn 1985): 290–299.

Draper, John William. *History of the Conflict Between Religion and Science*. New York: D. Appleton and Co, 1874.

Eng, David L, Judith Halberstam, and José Esteban Muñoz, eds. "What's Queer About Queer Studies Now?" Special Issue of *Social Text* 23 no. 3–4, 84–85 (Fall-Winter 2005).

Fausto-Sterling, Anne. *Sexing the Body: Gender, Politics and the Construction of Sexuality*. New York: Basic Books, 2000.

Foucault, Michel. *The History of Sexuality, 3 Vols*. New York: Random House, 1978–1986.

Gudorf, Christine E. "The Erosion of Sexual Dimorphism: Challenges to Religion and Religious Ethics." *Journal of the American Academy of Religion* 69 no. 5 (2001): 863–891.

Halperin , David M. *Saint Foucault: Towards a Gay Hagiography*. New York : Oxford University Press, 1995.

Haraway, Donna. *Simians, Cyborgs and Women: The Reinvention of Nature*. New York: Routledge, 1990.

Herdt, Gilbert, ed. *Third Sex, Third Gender: Beyond Sexual Dimorphism in Culture and History*. New York: Zone Books, 1993.

Jagose, Annamarie. *Queer Theory: An Introduction*. New York: University Press, 1997,

Keller, Catherine and Laurel Schneider, eds. *Polydoxy: Theology of Multiplicity and Relation*. New York: Routledge, 2011.

Knight, Douglas and Amy-Jill Levine. *The Meaning of the Bible: What the Jewish Scriptures and Christian Old Testament Can Teach Us*. New York: HarperCollins, 2011.

Levine, Amy-Jill, ed. *Feminist Companion to Paul: Authentic Pauline Writings*. New York: Continuum, 2004.

Lorde, Audre. "Age, Race, Class and Sex: Women Redefining Difference." *Sister Outsider: Essays and Speeches* (Freedom, CA: Crossing Press, 1984):114–123.

Malabou, Catherine. *What Should We Do with Our Brain?* New York: Fordham University Press, 2008.

Merchant, Carolyn. *The Death of Nature: Women, Ecology and the Scientific Revolution*. New York: HarperCollins, 1980.

Neill, James. *The Origins and Role of Same-Sex Relations in Human Societies*. Jefferson, NC: McFarland & Co, 2009.

Olsen, Richard. *Science and Religion, 1450–1900: From Copernicus to Darwin*. Baltimore, MD: Johns Hopkins University Press, 2004.

Roughgarden, Joan. *Evolution's Rainbow: Diversity, Gender and Sexuality in Nature and People*. Berkeley, CA: University of CA Press, 2004.

Ruether, Rosemary R. *Christianity and the Making of the Modern Family*. Boston, MA: Beacon Press, 2000.

Schippert, Claudia. "Queer Theory and the Study of Religion." *Revista de Estudos da Religião* 4 (2005): 90–99.

Sedgwick, Eve K. *Epistemology of the Closet*. Berkeley: University of CA Press, 1990.

Shipley, Maynard. *The War on Modern Science: A Short History of the Fundamentalist Attacks on Evolution and Modernism*. New York: Alfred A. Knopf, 1927.

Stenmark, Lisa. "Storytelling and Wicked Problems: Myths of the Absolute and Climate Change." *Zygon* 50, no. 4 (December 2015): 922–36.

———. *Religion, Science and Democracy: A Disputational Friendship*. Lanham, MD: Lexington, 2013.

White, Andrew. *A History of Warfare of Science and Theology in Christendom*. New York: D Appleton and Co., 1896.

White, Lynn. "The Historical Roots of Our Ecological Crisis." *Science* 155.3767 (1967): 1203–1207.

Whitehead, Alfred North. *Science and the Modern World*. New York: Macmillan, 1925.

Wilchins, Riki. *Queer Theory, Gender Theory: An Instant Primer*. Los Angeles: Alyson Books, 2004.

Wulf, Andrea. *The Invention of Nature: Alexander von Humboldt's New World*. New York: Vintage Books, 2015.

NOTES

1. See, for example, Lisa Stenmark, "Storytelling and Wicked Problems: Myths of the Absolute and Climate Change," *Zygon* 50, no. 4 (December 2015): 922–36.

2. Jacques Derrida, "Racism's Last Word," *Critical Inquiry* 12 (Autumn 1985): 292.

3. For more on the differences between them, see Michael Sprinker, "Textual Politics: Foucault and Derrida," *boundary 2* 8, no. 3 (Spring, 1980): 75–98.

4. Michel Foucault, *The History of Sexuality 3 vols* (New York: Random House 1978–1986).

5. The term "trans" was originally very broad, referring to anyone whose gender didn't "match" their sex—drag queens, butch women, and so forth. More recently, it has come to mean people who are transsexual.

6. Audre Lorde, "Age, Race, Class and Sex: Women Redefining Difference," in *Sister Outsider: Essays and Speeches* (Freedom, CA: Crossing Press, 1984), 114.

7. Ibid., 116.

8. Eve K. Sedgwick, *Epistemology of the Closet* (Berkeley: University of CA Press, 1990), 8.

9. Judith Butler, *Gender Trouble: Feminism and the Subversion of Identity* (New York: Routledge, 1990).

10. Riki Wilchins, *Queer Theory, Gender Theory: An Instant Primer* (Los Angeles: Alyson Books, 2004), 124.

11. Butler, *Gender Trouble,* 175–187.

12. Bruce Bagemihl, *Biological Exuberance: Animal Homosexuality and Natural Diversity* (New York: St. Martins Press, 1999); Joan Roughgarden, *Evolution's Rainbow: Diversity, Gender and Sexuality in Nature and People* (Berkeley: University of CA Press, 2004).

13. "Lesbian Lizards," *Battle of the Sexes in the Animal World*, on *BBC2,* www.bbc.co.uk/programmes/p006v48r. March 10, 2010 (1:45).

14. Anne Fausto-Sterling, *Sexing the Body: Gender Politics and the Construction of Sexuality* (New York: Basic Books, 2000), 46.

15. According to the Intersex Society of North America's statistics, the rates are even higher when one counts all of the different ways one's gender/sex can stray from the "norm." See, "How Common Is Intersex" on the website of the Intersex Society of North America: http://www.isna.org/faq/frequency. (Last accessed on August 9, 2017.)

16. Joan Roughgarden, *Evolution's Rainbow.*

17. On the human microbiome see: The National Institute of Health's Human Microbiome Project: http://www.hmpdacc.org/overview/about.php. (Last accessed on August 9, 2017.) On neuroplasticity see: Catherine Malabou, *What Should We Do with Our Brain?* (New York: Fordham University Press, 2008).

18. See, e.g.: Alfred North Whitehead, *Science and the Modern World* (New York: Macmillan, 1925).

19. Karen Barad, *Meeting the Universe Halfway: Quantum Physics and the Entanglement of Matter and Meaning* (Durham, NC: Duke University Press, 2007).

20. Barad, *Meeting the Universe Halfway*, 132–188.

21. Christine E. Gudorf, "The Erosion of Sexual Dimorphism: Challenges to Religion and Religious Ethics," *Journal of the American Academy of Religion* 69 no. 4 (2001): 863–891.

22. For a good history of the development of Victorian ideals about sex and sexuality see: Rosemary Radford Ruether, *Christianity and the Making of the Modern Family* (Boston, MA: Beacon, 2000).

23. Claudia Schippert, "Queer Theory and the Study of Religion," *Revista de Estudos da Religião* 4 (2005): 90–99.

24. John Boswell, *Same-Sex Unions in Premodern Europe* (New York: Vintage, 1994). See also Bernadette Brooten, *Love Between Women: Early Christian Responses to Female Homoeroticism* (Chicago, IL: University of Chicago Press, 1996).

25. See, e.g.: Daniel Biosvert and Jay Emerson Johnson, eds. *Queer Religion, 2 Vols* (Santa Barbara, CA: ABC-CLIO, 2012).

26. See, e.g.: Gilbert Herdt, ed., *Third Sex, Third Gender: Beyond Sexual Dimorphism in Culture and History* (New York: Zone Books, 1993).

27. See, e.g.: Douglas A. Knight and Amy-Jill Levine, *The Meaning of the Bible: What the Jewish Scripture and Christian Old Testament Can Teach Us* (New York: HarperCollins, 2011), 295–328.

28. See, e.g.: James Neill, *The Origins and Role of Same-Sex Relations in Human Societies* (Jefferson, NC: McFarland & Co, 2009).

29. See, e.g.: Amy-Jill Levine, ed. *Feminist Companion to Paul: Authentic Pauline Writings* (New York: Continuum, 2004).

30. Marcella Althaus-Reid, *Indecent Theology: Theological Perversions in Sex, Gender and Politics* (New York: Routledge, 2000).

31. Catherine Keller and Laurel Schneider, eds., *Polydoxy: Theology of Multiplicity and Relation* (New York: Routledge, 2011).

32. Gilles Deleuze and Felix Guattari, *A Thousand Plateaus: Capitalism and Schizophrenia* (Minneapolis, MN: University of Minnesota Press, 1987).

33. See, Whitney Bauman, "Religion, Science, and Nature: Shifts in Meaning on a Changing Planet," *Zygon: Journal of Religion and Science* 46 no. 4 (December 2011): 777–792.

34. For a rich discussion, see Carolyn Merchant, *The Death of Nature: Women, Ecology and the Scientific Revolution* (New York: HarperCollins, 1980).

35. On the importance of travel for building modern understandings of nature, see, e.g.: Andrea Wulf, *The Invention of Nature: Alexander von Humboldt's New World* (New York: Vintage Books, 2015).

36. Whitney Bauman, *Religion and Ecology: Developing a Planetary Ethic* (New York: Columbia University Press, 2014), 37–62.

37. John William Draper's *History of the Conflict Between Religion and Science* first appeared in 1874, and Andrew White's *A History of the Warfare of Science and Theology in Christendom* was published in 1896. Both books are widely available online, including Project Gutenberg and Google Books.

38. Maynard Shipley, *The War on Modern Science: A Short History of the Fundamentalist Attacks on Evolution and Modernism* (New York: Alfred A. Knopf, 1927), 3. For a fuller discussion, see Richard Olson, *Science and Religion, 1450–1900: From Copernicus to Darwin* (Baltimore: Johns Hopkins University Press, 2004), 5–6.

39. Ian G. Barbour, *When Science Meets Religion: Enemies, Strangers, or Partners?* (New York: HarperCollins, 2000).

40. Lynn White, "The Historical Roots of Our Ecological Crisis," *Science* 155. 3767(1967): 1203–1207.

41. You can read more about and actually take the Ecological Footprint Quiz here: http://www.footprintnetwork.org/our-work/ecological-footprint/. (Last accessed on August 9, 2017.)

42. Bauman, *Religion and Ecology*, 7.

43. Lisa Stenmark, *Religion, Science and Democracy: A Disputational Friendship* (Lanham, MD: Lexington, 2013).

44. The term nature-culture is taken from: Donna Haraway, *Simians, Cyborgs and Women: The Reinvention of Nature* (New York: Routledge, 1990).

45. David L. Eng, Judith Halberstam, and José Esteban Muñoz, eds. "What's Queer About Queer Studies Now?" a Special Issue of *Social Text* 84–85 23/3–4 (Fall-Winter 2005), 12.

46. Gloria Anzaldúa, *Borderlands/La Frontera: The New Mestiza* (San Francisco: Spinsters/Aunt Lute, 1987), 80, 81.

47. Annamarie Jagose, *Queer Theory: An Introduction* (New York: University Press, 1997), 1.

48. David M. Halperin, *Saint Foucault: Towards a Gay Hagiography* (New York: Oxford University Press, 1995), 62.

Chapter One

Both/And

Science, Religion, and the Fluidity of Identity

Philip Clayton and Kirianna Florez

Typically, authors in the science-religion field start with the dichotomy between science and religion, understood as foreign to each other: yin *versus* yang, male *versus* female. The authors then seek to reconcile them in one way or another, transforming warfare into harmony and creating sameness out of fundamental difference. But what if this entire approach, dominant in the science-religion field, is mistaken?

In this chapter, we focus on the binary of gender (male and female) and how an understanding of gender fluidity can create a new ethical framework. We begin by introducing the complexities of sex, gender, gender expression, and sexuality, which offer an expansive ontology. We explore expressions of gender fluidity within Western Christianity; we then reconsider science through the lens of gender fluidity, showing that applications of science do not rise above these complexities. Finally, we argue for using gender fluidity—as highlighted in our discussion of religion and science—as a model for an ethical space in which the dichotomous either/or, us/them gives way to multiple continua of lived experiences that command equal respect and empathy. In the end, gender fluidity—the fluid movement across sexual and gender differences that "cannot be crossed and must not be crossed"—becomes not only a model for one field but a core of a way of being in the world.

SEX AND GENDER

Sex, gender, and sexuality are aspects of human identity. The varieties of sex, gender, and sexuality within human experience, and their interrelationships, are massively complex.

Sex is determined by genetics and environment. Because humans are a sexually dimorphic species, different physical sex characteristics, or phenotypes, express the underlying genotypes. The sexual dimorphism found in humans is caused by genetic markers, which carry the blueprints that control the production of hormones within the body; for example, the XX genetic marker carries instructions for different hormone production than the XY genetic marker. Differences between sexes are manifested both in primary sex characteristics, such as genitalia and internal reproductive organs, and in secondary sex characteristics, such as breast growth and increased facial hair. These phenotypical differences are caused by hormone production, which is influenced by genetic markers and/or by the environment.

The most common genetic sexes for humans are XX-female and XY-male, but there are many other genetic possibilities that can determine sex, including XX-males, XY-females, XXY, XYY, XO, etc. In addition, genetic sex does not always match the physical sex characteristics that appear. For example, XX-males have the genetic marker XX, like XX-females, but XX-males' phenotypical characteristics are more in line with XY-male phenotypes, including XY-male genitalia, facial hair, and other features. In short, while sex is a genetically determined characteristic, the genetic sex does not always match the phenotypical sex of an individual. If an individual has ambiguous primary sex characteristics, or if an individual's genetic and phenotypical sex do not align, the person may be considered *intersex*. Note that the concept of intersex is a socially constructed category. Like all social and cultural categories, its semantic range and connotations are constantly being reevaluated and adapted.[1]

While biological sex is determined by genetics and hormones, gender more clearly expresses a variety of socially constructed roles. Socially constructed gender roles affect the divisions of labor within society, ranging from intimate household-based tasks such as childcare or food preparation to public tasks such as politics or religious leadership. Gender roles also affect how individuals are expected to perform in their individual lives, including areas such as dress, movement, language, and social interaction. When people's gender identity differs from their genetically determined sex—Joan's genetic and phenotypical sex, for example, is male, whereas her gender identity is female—they are considered to be *transgender*. The terms Assigned-Female-At-Birth (AFAB) and Assigned-Male-At-Birth (AMAB) more accurately identify sex and gender realities than the dichotomous terms *female* and *male*. An individual may have been born with the phenotypical sex characteristics of a female or male without being gendered in the same way.

These data suggest an emphatically non-dichotomous ontology: neither genetic sex nor gender can be understood within a two-part logic. Even in cultures where the most visible genders are masculine and feminine, actual gender includes feminine males, masculine females, individuals who are agender, nonbinary, or androgyne, etc. In addition, a person's gender includes both their gender identity and their gender expression. *Gender identity* is an individual person's experience of their gender, their "personal sense of identity as masculine or feminine, or some combination thereof."[2] But the person's *gender expression* may be masculine, feminine, both, neither, or multiple (fluid). Just as genetic sex and phenotypical sex may or may not align, gender identity and gender expression also may or may not align. For example, an individual who identifies herself as a woman may express herself through dress, language, and social roles which are socially considered to be masculine.

Finally, *sexuality or sexual orientation* is the sexual desire an individual person experiences. While the classification of sexual orientations is affected by sex and gender, sexual orientation is clearly not determined by either one. Another range of concepts must be used: sexually attracted to people of a different sex or gender (heterosexual); sexually attracted to people of the same sex or gender (homosexual); sexually attracted to people of the same sex or gender as well as different sexes or genders (bisexual or pansexual); not experiencing sexual attraction or desire (asexual); etc. There does not need to be any connection between a person's sex and gender and their sexuality or sexual orientation. For example, Rachel is an AFAB woman who is sexually attracted to women, and Samuel is an AFAB transman who is sexually attracted to women. While the genetic and/or phenotypical sex of both is female, and while both are sexually attracted to women, their sexual orientation will be classified differently because of their different genders: Rachel is homosexual because her gender is the same as the gender to which she is sexually attracted, and Samuel is heterosexual because his gender is different from the gender to which he is attracted.

The possible identities expand further. Instead of a single dichotomy, "male" *versus* "female," we now recognize no less than four different factors: genetic sex, gender, gender expression, and sexual preference. Each one offers at least four possibilities: male/masculine, female/feminine, both, neither. Picture these 16 possibilities listed along the top of a chart, and 16 possibilities listed down the side, representing the types of attraction to another person. That yields a staggering 256 different boxes or kinds of relationship. And yet this move from the traditional 2 to 256 possibilities is only the beginning. Each of the 16 factors is actually a continuum, not merely four discrete options within one of the four categories. Now add the indeterminacy of translation across cultural and subcultural systems: is a particular square on the chart in Mexican Spanish really the same as in Sanskrit? Or think of semantic differences between Toledo, Ohio and San Francisco—or, for that matter, subtle differences of meaning between 15-year-olds and 65-year-olds living in Berkeley. The options expand exponentially, deconstructing any attempt at a fixed system of boxes.

To any science that would restrict the categories to a discrete, limited set of options, this explosion of possible identities and relationships will appear off-putting, bizarre, or unnatural. More than just a non-dichotomous approach to human being and identity, we have a rather queer ontology.

GENDER AND QUEER STUDIES

Gender Studies, which emerged in the 1970s, is the interdisciplinary study of literature, history, art, anthropology, politics, and other fields in which sex and gender are placed at the forefront. Although Queer Studies arose within Gender Studies, we follow many other authors in the belief that it has significant implications that are not limited to the study of gender alone. As Susannah Cornwall notes, "the very concept of queer has built into it from the start an idea of elusiveness, uncertainty, non-fixity, and a resistance to closed definitions."[3] To be categorized as queer is to be placed outside the boundaries of acceptability, to be rendered abnormal. One can turn this exclusion into a place of resistance, giving it its own name and identity: we are the LGBTQ community and we oppose the injustices of heteronormativity. Note that to affirm this label of exclusion is *to embrace an identity of exclusion*, an act that locates us in a liminal place of contradiction. We deconstruct the act of being queered, but only at the cost of unsettling our own identity.

The definition of "queer," and with it queer theory, thus becomes uncertain and elusive—permanently unsettled. If queer means anything that is "at odds with the normal, the legitimate, the dominant,"[4] then both "queering" and "being queered" may begin to name a much broader social phenomenon, expanding from their original context in sexual politics. The terms are also dynamic: actions and identities that are now viewed as normal, legitimate, and acceptable may in the past have been abnormal, illegitimate, and deviant. Queer may become unqueer, and the unqueer, queer.

Two more implications follow. When "queer" labels whatever is not normal or legitimate, or whatever is excluded by the dominant group, then it can become a transferable badge of honor, worn by other persons or groups. When anything and everything can be queer, then the significance of the term is lost. But the opposite can also occur when queerness and queer theory come to be viewed through too narrow a lens. Cornwall notes that this is a particular problem within queer theology, which other theologians tend to treat as if it were only concerned with sexuality. Although queer theology in fact means multiple things to multiple theologians—from sex to resistance to deconstruction—its semantic range is often lost. This leads to what Cornwall calls the "image problem": "the assumption or *perception* that queer theology is just to do with sex or, even more narrowly, just to do with homosexuality . . . may present a potential obstacle to its utility for theologians."[5] This foreshortening to sexuality alone ties the term so tightly to one specific location that its queerness is lost altogether.

In short, queer theory can be understood both narrowly and broadly. A narrow reading of queer theory focuses on issues of sex, gender, and sexuality, while a broad reading of queer theory takes into consideration anything that transgresses boundaries. Narrow and broad readings of queer theory are not mutually exclusive; they can and do affect each other.

Discussions of science and theology have much to learn from the exploration in these last pages. Gender fluidity is not only an ontology, as we argued above, but also an *epistemology*, a way of approaching the world. It spawns an "abnormal" and "warped" analysis of established fields of study and, we will argue, new forms of caring. In an article on lesbian/gay and queer theology, Mary Elise Lowe noted that queer theology had borrowed six insights from queer theory: "the method of deconstruction, the assertion that all meaning is constructed, the insight that gender is performed, the claim that identity is unstable, and the commitment that persons are constituted by discourses and subjected in the process."[6] In the following pages, we will engage two of these insights in particular: the concept of gender as performance, and the belief in the instability of identity.

GENDER FLUIDITY IN RELIGION

Gender emerges as a fluid category in religious writing in the West. It is and has been assigned to and removed from religious figures and concepts, with ramifications for theology and for relations between humans and the divine. Some of the clearest examples of gender fluidity can be seen in relation to the figure of Jesus Christ, as portrayed by authors ranging from theologians to poets. In this section we look at several examples of gender fluidity, mostly drawn from Western Christianity: the Bride figure of the New Testament and its potential identification with the Church; gender fluidity in John Donne's religious poetry and sermons; a contemporary Christian movement that encourages gender fluidity in role play based on Bridal Theology; and the fluid genders of deific figures, with examples of the figure of God in the Hebrew Bible.

In several books of the New Testament, a figure called the "Bride" is mentioned. Though the Bride works as a counterpoint to the Jesus-as-Bridegroom motif that appears in the Gospels, the identification of the Bride has never been concretely determined, thus leaving it open to interpretation. In the Book of Revelation, New Jerusalem is identified as a bride, for example in Revelation 21:2, where New Jerusalem appears "prepared as a bride adorned for her husband." This appearance of New Jerusalem as a bride follows the motif of identifying Jerusalem, Judah, and Israel as the bride(s) of God in the Hebrew Bible, as seen in Hosea, Deutero-Isaiah, Jeremiah, and Ezekiel. Another potential identity for the Bride can be found in Ephesians 5:22–23, which discusses the parallel relationships of husband-wife and Christ-Church, implying that the Church fills the role of the Bride within the Jesus-as-Bridegroom motif. This interpretation of Church-as-Bride and Jesus-as-Bridegroom can also be found in 2 Corinthians 11:2–4, in which Paul admonishes the Church in Corinth, stating that he "promised [the Corinthian Church] in marriage to one husband, to present [the Church] as a chaste virgin to Christ." These latter texts, which portray the Church as the Bride to Jesus-as-Bridegroom, influenced *bridal theology*, which views the relationship between Christ and the Church as a husband-and-wife relationship. Following the structure of a husband-and-wife relationship, Christ is viewed as the male spouse and the Church—including the male followers of Christ—are viewed as the female spouse(s).

The 17th-century theologian and poet John Donne gendered both himself and his soul as female in different religious and devotional texts. An example of Donne's fluidity of gender in poetry can be found in his Holy Sonnet X, "Batter my Heart," in which Donne uses two different types of female-based imagery to explore his relationship to God. First he describes himself "like an usurp'd town" that desires to be overtaken by God. Towns—like land—are characterized as female, and Donne's self-identification as an "usurp'd town" genders him as female and as "Labor[ing] t'admit [God]." By the end of the sonnet, Donne has again gendered himself as a female-bodied person who has been "betroth'd unto [God's] enemy," and who desires God to "Take me to you, imprison me, for I, / Except you enthrall me, never shall be free, / Nor ever chaste except you ravish me."[7] Donne's love and desire for God is expressed in gendered terms, with Donne imagining himself as feminine in his relationship with God.

Donne's gender-fluid literary identity appears also in his sermons, where he repeatedly genders his spirit as feminine and imagines himself—and other male Christians—as female partners in a marriage with God. A prime example is found in Donne's Washington wedding sermon, preached on May 30, 1621. Within this sermon Donne explores the parallels between secular marriage, spiritual marriage, and eternal marriage. While discussing the spiritual marriage between Christ and the soul, Donne describes his soul as the female captive who may be married to her captor, stating that he can "see [Christ's] face in all his temporall [sic] blessings, having shaved her [Donne's soul] head in abating her pride, and pared her nails in contracting her greedy desires, and changed her clothes not to fashion her self after this world, my soul being thus fitted by himself, Christ Jesus hath maried [sic] my soul."[8] Imagining himself as gendered female, performing female gender and sexual roles, allowed Donne to more deeply explore the relationship between humans and the divine.

A contemporary example of the fluidity of gender in Christian thought and theology can be found in the Family International, a modern denomination that began in the 1960s. Within Family International, teenagers and adults are encouraged to imagine themselves as female-bodied lovers of Jesus while they have sex or masturbate. This imaginative role play is meant to strengthen the loving relationship between the individual members of the church and Jesus; it is encouraged for all members, regardless of their own sex and gender. This practice stems from the Family International's understanding and usage of bridal theology, as they "believe that the marital metaphor used in the Bible to describe the intimate spiritual relationship between Jesus and His Church is meant to represent the passionate union of heart, mind, and spirit that Jesus seeks with each of His followers."[9] We must note that the Family International denomination only recognizes heterosexuality as a valid sexuality. The concern for avoiding the appearance of homosexuality within the denomination likely plays a role in the encouragement for men to view themselves as women/female-bodied when imagining sexual acts with Jesus. Though the Family International is not supportive of the LGBTQ community, we have including them in this discussion to highlight the way gender is viewed as (at least partially) fluid within contemporary Christian practices, even non-LGBTQ-friendly Christian communities.[10]

Gender fluidity also occurs around deific figures. One example is the fluidity of God's gender within the Hebrew Bible, which can be found in the names used for God. Within the Hebrew Bible God has a variety of names, including Shaddai, which also appears seven times as El Shaddai. The term (El) Shaddai is commonly translated into English as "(God) Almighty." A more literal translation of the Hebrew term, though, may be either "God of Mountains" or "God of Breasts." The term *el* means "god"; the term *shaddai*, however, has several possible definitions, allowing a wide range of interpretation. One possible etymological link for *shaddai* may be the Akkadian term *šadum*, which means "mountain." A second possibility is the Hebrew term *shad*, which means "breasts"; the Hebrew *shad* may also be connected to the Akkadian *šadum*. This second translation is particularly interesting, since the term "breasts" adds an element of fluidity to God's gender presentation and performance of gender roles.[11]

Additionally, the name El Shaddai appears in the Hebrew Bible seven times, with five occurences in Genesis and one in Exodus. Of these six, five are linked to blessings of fertility. The term Shaddai (without the accompanying El) is also used in Genesis 49:25, during Jacob's blessings for his sons. In this verse, Jacob states, " . . . from the God of your father, who will help you, and *Shaddai*, who will bless you, blessings of heavens from above and blessings of the deep lying below, and blessings of the breasts and the womb." The blessings linked to the term Shaddai include sexually female imagery such as Shaddai's breasts and womb. In short, within the Hebrew Bible God can be and is portrayed with female sexual characteristics and as playing female gender roles.

These examples—from the use of gender in the Church-as-Bride and Jesus-as-Bridegroom motif to the fluidity of God's presented gender and sex in the Hebrew Bible—show how fluid gender and gender roles often are within religion, in this case Judaism and Christianity. When we read and imagine these roles in religious practice, we are reminded of the two key aspects of Queer Studies that Mary Elise Lowe identifies: understanding gender as performance, and recognizing the instability of identity. In the above examples, gender is performed in theological writings of John Donne and when men and women imagine themselves as the female-bodied lover of Christ during intercourse and masturbation. Similarly, the instability of identity is not only human; the identity of the Hebrew God also shifts in different contexts, expressed through the different names that the canonical authors used. Freed from sexual and gender dichotomies, our identities flow freely into queer places. Even the very male/masculine God of the Torah grows breasts and opens her fertile womb.

SCIENCE AS A COUNTEREXAMPLE?

The religious examples we have explored show the fluid nature of gender and sexual self-understandings on the part of believers, a fluidity that manifests also in their ideas of the divine. Is science queer in similar ways?

At first blush science looks like a radically different case. Of course, there are scientists who are gender fluid, and many clearly affirm their own queer identities. But the scientific community as such views the lifestyles and self-conceptions of individual scientists as incidental to their work, in the same way that race, gender, socio-economic status, class, or nationality are viewed as incidental.[12]

Consider the contrast. "Religion" is notoriously difficult to define; universal standards imposed by scholars or members of a particular tradition beg the question for individual communities who dispute these boundaries. Categories of religious belief and practice are queer in the broader sense of the term, since opposition to norms, transformation of identities at the individual and corporate level, hybridity and syncretism—in short, every conceivable form of boundary crossing—are visible in a vast variety of religious phenomena. By contrast, since at least Francis Bacon (1561–1626), scientists have defined science and scientific success in rigid terms. Verifiability, falsifiability, replicability, the centrality of natural laws and quantitative reasoning, the testability of theories by data, and above all the objectivity of this process—these are the features that, in one way or another, have contributed to and continue to define the present practice of scientists working across the scientific disciplines.

Of course, one can stand outside science and say that these should not be the norms, the required conditions for something to count as science. Clearly there are gray areas, just as there are numerous cases of scientists breaking these standards.[13] On the one hand, there are no a priori reasons why the word "science" cannot be given different meanings. Culturally and historically there have been many different alternatives: medieval science (*scientia*), science as philosophy as in German idealism (*Wissenschaft*), Taoist science,[14] Buddhist science,[15] and many segments of classical Indian thought, such as the science of reincarnation. On the other hand, over the last few centuries, science has become a clearly defined global institution. Acknowledgments of the nature and value of the scientific enterprise have been made by significant numbers, and perhaps even majorities, of people from every culture of the world. Interestingly, even advocates of alternative ways of knowing the world use empirical science as the standard, arguing that their activities meet these standards.[16]

The institution of science as such is not gender fluid, but could it be fluid in a broader sense? It does, after all, challenge boundaries that are part of common sense and the everyday realities of most human beings: whether mass is solid, whether humans are qualitatively distinct from other animals, whether values exist, whether actions are free, whether God exists. But when it comes to questions of what counts as science, one generally does not find fluid definitions. Leading scientists have resisted the notion that quantum phenomena are indeterminate, that other mammals have culture, that incredibly small "strings" could be the foundations of all reality, or that our universe is part of a massively larger multiverse. Yet there has been no wavering among scientists on what is required; these theories will count as scientific only if they make empirically testable predictions that, when actually tested, can be verified by scientists around the world.

Significantly, scientists have discovered and continue to study many cases of sexual and gender fluidity. The traditional male-female dichotomy in biology was based on the size difference between gametes; small gametes (sperm) are produced by males and large ones (eggs) by females. But the groundbreaking work of the Stanford biologist Joan Roughgarden has undercut many of the traditional sexual boundaries. Her studies of clown fish, for example, trace the change of sex from male to female and back. "If [the female] is removed, the remaining male turns into a female, and one of the juveniles matures into a male."[17] Once the assumption of fixed sexual categories was removed, a mass of new data began to emerge. In different species sexes change simultaneously, or sequentially, or they crisscross back and forth. The widespread belief that sexual and gender categories are fixed in biological organisms turns out to the false; nature is far queerer than we imagined.

But the queering of the natural world is not the same as the queering of science. In fact, the brilliant work that Prof. Roughgarden and others have done has been taken as a sign of the importance of the scientific method as traditionally defined and practiced. One final point: the fact that Joan Roughgarden is a transgendered woman scientist has clearly influenced the phenomena that she studies, as she herself emphasizes. But to the scientific community, this fact does not queer science. Prof. Roughgarden's conclusions about clown fish are or are not verified by the data (in fact, they have been). By contrast, both she and the broader scientific community agree that the implications that she draws from her work for ethics, politics, and worldview are not themselves part of the science.

QUEERING THE INTERPRETATION OF SCIENCE: THE CRUCIAL ROLE OF THE RELIGION-SCIENCE DISCOURSE

When it comes to interpreting and applying science, however, the sharp dichotomy between science and non-science quickly collapses. When scientific results are interpreted philosophically, scientific criteria no longer control the process. Scientists are no more likely to be authorities on religious matters than football players are authorities on toothpaste or deodorant.

Scientists, in short, do not stand above the fluidity of interpretation; they swim in it, just like the rest of us. And yet even the craziest non sequiturs that they claim to derive from their work continue to have an immense influence on culture and politics, and indeed on the worldview that has come to dominate the modern period. Science, many scientists tell us, has proven *materialism* (only matter/energy exists, and all beliefs in spiritual realities are false), *reductionism* (all more complex phenomena, organisms, and ideas, however much they may appear to be real and to have effects in the world, are reducible to lower-level objects and laws), *determinism* rather than freedom (neurons, genes, and, ultimately, sub-atomic particles and forces cause all phenomena to be what they are), and *relativism* (nature is value-free, so human existence must likewise be).

Particularly disturbing is the dogmatic certainty with which these alleged implications of science are proclaimed by its self-designated spokespersons, such as the former professor "for the public understanding of science" at Oxford University, Richard Dawkins. Their famous quips represent dichotomous thinking at its most blatant:

> The more the universe seems comprehensible, the more it also seems pointless.
>
> —Steven Weinberg[18]

> Reductionism, given its unbroken string of successes during the next three centuries, may seem today the obvious best way to have constructed knowledge of the physical world, but it was not so easy to grasp at the dawn of science.
>
> —E. O. Wilson[19]

> Wires and chemicals, that's all we are, wires and chemicals.
>
> —Allan Basbaum[20]

> "You," your joys and your sorrows, your memories and your ambitions, your sense of personal identity and free will, are in fact no more than the behavior of a vast assembly of nerve cells and their associated molecules. As Lewis Carroll's Alice might have phrased it, "You're nothing but a pack of neurons."
>
> —Francis Crick[21]

These scientists are the most dichotomous of all when it comes to the relationship between science and religion:

> The truth . . . is that the conflict between religion and science is unavoidable. The success of science often comes at the expense of religious dogma; the maintenance of religious dogma *always* comes at the expense of science.
>
> —Sam Harris[22]

> Violence, irrational, intolerant, allied to racism and tribalism and bigotry, invested in ignorance and hostile to free inquiry, contemptuous of women and coercive toward children: organized religion ought to have a great deal on its conscience."
>
> —Christopher Hitchens[23]

> Faith is the great cop-out, the great excuse to evade the need to think and evaluate evidence. Faith is belief in spite of, even perhaps because of, the lack of evidence.
>
> —Richard Dawkins[24]

This dogmatism parading as science has been one of the main inspirations for the rapid growth of studies on science and religion. Scholars in this new field have emphasized the immense social damage caused by these attacks on the unique features of human existence, values, spirituality, and the sacred. But their research has also established the lack of justification for claims that seek to dichotomize and exclude. Science as such does not by itself settle claims in all other domains of human existence. Generalizations from science to "reality as a whole" simply ignore the complex webs of cultural, linguistic, and historical identities. Thomas Hobbes wrote in 1648, for example, that everything is merely "matter in motion," from which he derived the value judgment that life is "nasty, brutish, and short." Yet from the fact that a given science can predict the dynamics of matter in motion it does not follow that material particles are the only form of reality.

At first it might seem as though the "science only" dogmatists are among the most queer (in the broad sense of the term) of all, because they cross standard boundaries and redefine them as they go. But to cross the boundaries of what is normally considered acceptable without acknowledging one's own queerness in the process is an act of hegemony and violence, a move to eliminate the difference of the other. The contribution of studies of science and religion is to bring this dynamic to the surface. In what ways does scientific practice come to function in religious ways when it is transformed from method to worldview? In what ways are the supposedly neutral scientists functioning more as "true believers"?[25] Have scientists become the high priests of our day, and has technology become the Holy of Holies, hidden behind the curtain so that only the high priests truly understand it?

It is no coincidence that the study of religion and science would play a unique role in this deconstruction of the universalizing claims of these scientists. After all, it is the scientific spokespersons themselves who have castigated religious beliefs as the queerest and religious practices as the most disgusting and most taboo of all. Religion-science scholars have been able to pull back the curtain, revealing the ones who are turning the dials and pulling the levers to create the universal Wizard of Oz. At first it may have appeared that the universalizers of science were the ones most adept at challenging structures and making categories fluid. In the end, however, it turns out that they have treated their own categories as the most inviolate, whereas it is often (but not always) the religious believers who are bending categories and moving fluidly around and within them. The implications of science for the rest of human existence are not monolithic; they are subtle, changing, ideally life-giving and transformative.

To see this point, we close this section by returning to the broader work of Joan Roughgarden as a scholar in Gender Studies, which is *not* a value-neutral discipline. The studies published by Roughgarden reveal the gender fluidity in nature itself. As she emphasizes, "male and female functions don't need to be packaged into lifelong distinct bodies. Hermaphroditic vertebrate species are successful and common."[26] Interpreting these results through insights from human gender studies transforms our view of the world around us, and thus also of ourselves:

> "Gender" usually refers to the way a person expresses sexual identity in a cultural context. Gender reflects both the individual reaching out to cultural norms and society imposing its expectations on the individual. Gender is usually thought to be uniquely human—any species has sexes, but only people have genders. With your permission, though, I'd like to widen the meaning of gender to refer to nonhuman species as well. As a definition, I suggest: *Gender is the appearance, behavior, and life history of a sexed body.* A body becomes "sexed" when classified with respect to the size of the gametes produced. Thus, gender is appearance plus action, how an organism uses morphology, including color and shape, plus behavior to carry out a sexual role.[27]

The results have deep ethical and political implications. As Roughgarden writes,

> Scientists must start to teach the truth, and organizations who present nature to the public must start offering an accurate picture. Scientists are professionally responsible for refuting claims that homosexuality is unnatural. The dereliction of this responsibility has caused homosexual people to suffer persecution as a result of a false premise of "unnaturalness," and to suffer low self-worth and personal dignity. Suppressing the full story of gender and sexuality denies diverse people their right to feel at one with nature and relegates conservation to a niche movement—the politics of a privileged identity.[28]

Here the acknowledgment of gender and sexual fluidity has significant social and political implications, just as the denial of fluidity in the name of static dichotomous concepts is damaging to those whose lived existence moves around and through and between the categories. The fluidity of gender has much to teach us about the fluidity of categories in general. Queerness in the specific sense of sexual and gender fluidity opens our eyes to queerness in the broad sense. We can now seek and find a place beyond the either/or world of male/female, spurned/embraced, in/out, right/wrong.

SEX/GENDER FLUIDITY AS WORLDVIEW: TOWARD A QUEER ETHICS

What is my sex, my gender, my gender expression? Whom do I long for as a partner? These questions concern fundamental ways of being in the world. In recent decades feminism, Gender Studies, and now queer theory have radically shifted the understanding of how the four questions interact. Widely held beliefs notwithstanding, the debate is not just about "sexual orientation"; it concerns one's entire life-orientation. As we have seen, these questions touch on what is and how we know it—ontology and epistemology. They cross boundaries into religion, science, and other fields. They also open the door to a distinctive approach to ethics.

Studies of sex and gender have undercut dichotomous thinking, but Queer Studies has not always welcomed the resulting fluidity. Some respond by bifurcating into insiders and allies, with the message to allies that "you can help us, but you are not us." Others erase the differences altogether: "we are all queer now," without realizing that, when everyone is queer, no one is.

There is an alternative, however, which we will introduce in narrative form. Consider John, a WASP (white Anglo-Saxon Protestant) heterosexual man who lives in an upper middle-class American suburb. Imagine him not overly reflective or empathetic. If John views life through a dichotomous lens—either/or, us/them, yes/no—then he will view people who are different from him as "others." Any number of things can separate him from these "others": gender, sexuality, race, religion, language, country of origin, economic class, etc. Locked in these either/or dichotomies, John will have difficulty viewing "the others" as people who should receive the same respect and empathy that he has for himself and those like him.

Also, bifurcated in this way, John will be forced into the same deconstructive dilemma just identified. On the one hand, he may relate to some of "those others" by dividing people into insiders and allies. Where he is not "one of them," he can at best become a helpful ally to "them," helping "them" from the outside. (Of course, he can also become their enemy and adversary.) On the other hand, he may declare himself part of the group: "We are all immigrants here," or "All the religions of the world are one, so all of us are brothers and sisters." This over-generalization has the very real danger of erasing unique experiences and silencing the voices of the underprivileged. Given only these two options, John will either essentialize each difference or erase it altogether. With these two options, either some people are sexually queer and John helps "folks like them" from the outside, or everyone is queer now, which denies the unique experiences of the LGBTQ community. Neither response is ideal.

We advocate a new paradigm. One can shift from an either/or, yes/no worldview to a worldview of continua, in which individuals—regardless of their differences—share certain common experiences of personhood. If John adopts this paradigm of fluidity, he learns to imagine himself in the position of his neighbors. Just as religious rituals may prompt individuals to imagine themselves as a different gender, John can imagine himself living life as a different person in sex, gender, gender expression, or sexual attraction—or, for that matter, as having a different race, culture, language, or socio-economic identity. Imagining the lived reality of one's neighbors opens the door to a both/and identity; one recognizes shared experiences *and also* acknowledges real differences. The practice of imagination makes it possible to begin experiencing multiple sides of what were once strict dichotomies. Imagination births empathy and understanding; together, they grow a new ethical space.

It is important to note that this paradigm of fluidity does not change John's actual lived reality. John remains a WASP, wealthy, heterosexual man; he will not become a person of color, nor will he become LGBTQ. His lived reality and that of his neighbors will remain different. However, John can acknowledge that he lives on a continuum of sameness and difference, and that he may also be "othered" by people who view him as a bit queer not because of his sexual orientation or gender expression, but for other reasons. He can acknowledge that his lived reality is only one possibility, recognizing the legitimacy and importance of lived realities that are different from his own. By imagining himself living his neighbors' lives, John will recognize his neighbors' humanity, and therefore treat each of his neighbors with the same respect and care as John expects for himself, regardless of his neighbors' physical and/or social location. By recognizing the fluidity of his own identity, as well as the fluidity of the identities of other persons, John can enter into the space of a queer ethics.

CONCLUSION

As we have seen, the queering of religion and science is a world- and life-view. As we study them, the worlds of science and religion turn out to be far less rigid and far more fluid than many acknowledge. Sometimes gender fluidity appears explicitly in the phenomena that science studies or in religious beliefs and practices; at other times, as we have seen, it is in the applications of discoveries and the living out of beliefs that traditional dichotomies collapse. Like the rest of nature, the expression of our humanity is complex, multi-factorial, and as far from bifurcated as one can imagine.

In these pages we have sought to evoke and to reflect upon the new paradigms that seek to comprehend this shift. The queering and fluidity of gender within both science and religion expresses a non-dichotomous worldview. They engender a worldview not of dichotomies but of continua. This paradigm becomes revolutionary when one considers its potential to transform questions of identity and justice, whether social, racial, or ecological.

WORKS CITED

Baker, Eileen. "From the Children of God to the Family International: A Story of Radical Christianity and De-radicalising Transformation." In *Handbook of Global Contemporary Christianity: Movements, Institutions, and Allegiance,* vol. 12. Edited by Stephen J. Hunt. Leiden: Brill, 2016. 402–21.

Biale, David. "The God with Breasts: El Shaddai in the Bible." *History of Religions,* 21, no. 3 (February, 1982): 240–56.

"The Bride of Christ." *The Family International.* Accessed February 26, 2017. http://www.thefamilyinternational.org/en/about/our-beliefs/bride-christ/.

Bronowski, Jacob. *Science and Human Values.* New York: Harper & Row, 1956.

Capra, Fritjof. *The Tao of Physics: An Exploration of the Parallels between Modern Physics and Eastern Mysticism.* Boston: Shambhala, 2010.

Cornwall, Susannah. *Controversies in Queer Theology.* London: SCM Press, 2011.

Crick, Francis. *The Astonishing Hypothesis: The Scientific Search for the Soul.* London: Simon and Schuster, 1994.

Dawkins, Richard. Speech to the Edinburgh International Science Festival, April 15, 1992.

Donne, John. "Holy Sonnet X: Batter my heart." In *John Donne's Poetry.* Edited by Donald R. Dickson. New York: W. W. Norton & Company, Inc., 2007.

———. "Preached at a Marriage [the marriage of Mistress Margaret Washington at the church of St. Clement Danes, May 30, 1621]." In *The Sermons of John Donne* 3, no. 11. Edited by George Reuben Potter and Evelyn Mary Spearing Simpson. Berkeley: University of California Press, 1957. Accessed March 29, 2017. http://contentdm.lib.byu.edu/cdm/compoundobject/collection/JohnDonne/id/3161/rec/1.

Ferris, P. J. and U. W. Goodenough. "Transcriptions of novel genes, including a gene linked to mating-type, induced by *Chlamydomonas* fertilization." *Molecular Cellular Biology* 7 (1987): 2360–66.

Ferris, P. J., C. Pavlovic, S. Fabry, and U. W. Goodenough. "Rapid Evolution of Sex-related Genes in *Chlamydomonas*." *Proceedings of the National Academy of Sciences* 94 (1997): 8634–39.

Gould, Stephen J. *The Mismeasure of Man*. New York: Norton, 1996.

Halperin, David M. *Saint Foucault: Towards a Gay Hagiography*. New York: Oxford University Press, 1995.

Harris, Sam. "Science Must Destroy Religion." *Huffington Post*, Jan. 2, 2006, updated May 25, 2011; http://www.huffingtonpost.com/sam-harris/science-must-destroy-reli_b_13153.html

Hitchens, Christopher. *God Is Not Great: How Religion Poisons Everything*. New York: Twelve [Hachette, Warner], 2007.

Kurian, George Thomas and Mark A. Lamport, editors. "Children of God." In *Encyclopedia of Christianity in the United States*, vol. 2. Lanham, MD: Rowman & Littlefield, 2016. 727.

Lowe, Mary Elise. "Gay, Lesbian, and Queer Theologies: Origins, Contributions, and Challenges." *Dialog* 48, no. 1 (March 2009): 49–61.

Lutzky, Harriet. "Shadday as a Goddess Epithet." *Vetus Testamentum* 48, no. 1 (January, 1998): 15–36.

Morrow, Deana F. "Sexual Orientation and Gender Identity Expression." In *Sexual Orientation and Gender Expression in Social Work Practice: Working with Gay, Lesbian, Bisexual, and Transgender People*. Edited by Deana F. Morrow and Lori Messinger. New York: Columbia University Press, 2006. 3–17.

Raymo, Chet. *Skeptics and True Believers: The Exhilarating Connection Between Science and Spirituality*. New York: Walker Books, 1998.

Roughgarden, Joan. *Evolution's Rainbow: Diversity, Gender, and Sexuality in Nature and People*. Berkeley: University of California Press, 2004.

Torrance, T. F. *Theological Science*. Oxford: Oxford University Press, 1978.

Wallace, B. Allan. *The Taboo of Subjectivity: Toward a New Science of Consciousness*. Oxford: Oxford University Press, 2000.

———, ed. *Buddhism and Science: Breaking New Ground*. New York: Columbia University Press, 2003.

Weinberg, Steven. *The First Three Minutes: A Modern View of the Origin of the Universe*, 2nd ed. New York: BasicBooks, 1993.

"What Is Intersex?" *Intersex Society of North America*. Accessed February 26, 2017. www.isna.org/faq/what_is_intersex.

Wilson, Edward O. *Consilience: The Unity of Knowledge*. New York: Knopf/Random House, 1998.

NOTES

1. "What Is Intersex?" *Intersex Society of North America*, accessed February, 2017, www.isna.org/faq/what_is_intersex.

2. Deana F. Morrow, "Sexual Orientation and Gender Identity Expression," *Sexual Orientation and Gender Expression in Social Work Practice: Working with*

Gay, Lesbian, Bisexual, and Transgender People, ed. Deana F. Morrow and Lori Messinger (New York: Columbia University Press, 2006), 8.

3. Susannah Cornwall, *Controversies in Queer Theology* (London: SCM Press, 2011), 9.

4. David M. Halperin, *Saint Foucault: Towards a Gay Hagiography* (New York: Oxford University Press, 1995), 62, quoted in Susannah Cornwall, *Controversies in Queer Theology*, 12.

5. Cornwall, *Controversies in Queer Theology*, 26.

6. Mary Elise Lowe, "Gay, Lesbian, and Queer Theologies: Origins, Contributions, and Challenges," *Dialog* 48 no. 1 (March 2009), 52, quoted in Susannah Cornwall, *Controversies in Queer Theology*, 27.

7. John Donne, "Holy Sonnet X: Batter my Heart," in *John Donne's Poetry*, ed. Donald R. Dickson (New York: W. W. Norton & Company, Inc., 2007), 140.

8. John Donne, "Preached at a Marriage [the marriage of Mistress Margaret Washington at the church of St. Clement Danes, May 30, 1621]," in *The Sermons of John Donne*, vol. 3, no. 11, ed. George Reuben Potter and Evelyn Mary Spearing Simpson (Berkeley: University of California Press, 1957), http://contentdm.lib.byu.edu/cdm/compoundobject/collection/JohnDonne/id/3161/rec/1

9. "The Bride of Christ," *The Family International*, accessed February 2017, http://www.thefamilyinternational.org/en/about/our-beliefs/bride-christ/.

10. George Thomas Kurian and Mark A. Lamport, editors, "Children of God," *Encyclopedia of Christianity in the United States,* vol. 2 (Lanham, MD: Rowman & Littlefield, 2016), 727.

For a more comprehensive survey of the Family International, including allegations of sexual abuse within the Family International, see Eileen Baker's "From the Children of God to the Family International: A Story of Radical Christianity and De-radicalising Transformation," *Handbook of Global Contemporary Christianity: Movements, Institutions, and Allegiance,* vol. 12, ed. Stephen J. Hunt (Leiden: Brill, 2016), 402–21.

11. David Biale, "The God with Breasts: El Shaddai in the Bible," *History of Religions,* 21.3 (February, 1982), 240–56; Harriet Lutzky, "Shadday as a Goddess Epithet," *Vetus Testamentum,* 48, no. 1 (January, 1998), 15–36.

12. Jacob Bronowski, *Science and Human Values* (New York: Harper & Row, 1956), especially chapter 1.

13. E.g., among many examples, Stephen J. Gould, *The Mismeasure of Man* (New York: Norton, 1996).

14. Fritjof Capra, *The Tao of Physics: An Exploration of the Parallels between Modern Physics and Eastern Mysticism* (Boston: Shambhala, 2010).

15. B. Allan Wallace, *The Taboo of Subjectivity: Toward a New Science of Consciousness* (Oxford: Oxford University Press, 2000); Wallace, ed., *Buddhism and Science: Breaking New Ground* (New York: Columbia University Press, 2003).

16. T. F. Torrance, *Theological Science* (Oxford: Oxford University Press, 1978).

17. Joan Roughgarden, *Evolution's Rainbow: Diversity, Gender, and Sexuality in Nature and People* (Berkeley: University of California Press, 2004), 33. See also the important work of Ursula Goodenough on sexual and gender transitions in

microorganisms, e.g., P. J. Ferris and U. W. Goodenough, "Transcriptions of novel genes, including a gene linked to mating-type, induced by *Chlamydomonas* fertilization," *Molecular and Cellular Biology* 7 (1987): 2360–66; P. J. Ferris, C. Pavlovic, S. Fabry, and U. W. Goodenough, "Rapid Evolution of Sex-related Genes in *Chlamydomonas*," *Proceedings of the National Academy of Sciences* 94 (1997): 8634–39.

18. Steven Weinberg, *The First Three Minutes:* A *Modern View of the Origin of the Universe*, 2nd ed. (New York: BasicBooks, 1993), Epilogue, 154.

19. Edward O. Wilson, *Consilience: The Unity of Knowledge* (New York: Knopf/ Random House, 1998), 33.

20. Public comment made at a neuroscience conference in Berkeley in 1999.

21. Francis Crick, *The Astonishing Hypothesis: The Scientific Search for the Soul* (London: Simon and Schuster, 1994), 3.

22. Sam Harris, "Science Must Destroy Religion," *Huffington Post,* Jan. 2, 2006, updated May 25, 2011; http://www.huffingtonpost.com/sam-harris/science-must-destroy-reli_b_13153.html

23. Christopher Hitchens, *God Is Not Great: How Religion Poisons Everything* (New York: Twelve [Hachette, Warner], 2007), 56.

24. Richard Dawkins, speech to the Edinburgh International Science Festival, April 15, 1992.

25. Chet Raymo, *Skeptics and True Believers: The Exhilarating Connection Between Science and Spirituality* (Walker Books, 1998).

26. Roughgarden, *Evolution's Rainbow*, 35.

27. Roughgarden, *Evolution's Rainbow*, 27.

28. Roughgarden, *Evolution's Rainbow*, 128.

Chapter Two

If You Quare It You Can Change It

Changing the Boxes That Bind Us

Emilie M. Townes

As a daughter of a molecular biologist who was also a faithful tither to her church, I grew up in a world that saw no clear separation between science and religion. It was in the air I breathed and the way in which I learned to see the world. For my mother, scientific exploration could only take her so far as she explored "The involvement of pH, Adenosine triphosphate, calcium, and magnesium in the contraction of the glycerinated stalks of Vorticella"[1] and other "decidedly microscopic" animals in her muscular dystrophy research in the mid-1950s to mid-1960s. "Then," she said, "it's all mystery and God."

The beauty of science was something that I was attracted to as a young child—particularly biology. I was fascinated, and remain so, that there is matter that is less than a cell. The flow of blood in our veins, photosynthesis in plants, stars in the sky, how a tornado forms, the power of a hurricane, the teeming world in our oceans, and more were interwoven with Sunday school lessons, tedious sermons, choirs who made a joyful noise to the Lord (sometimes literally), the many ways we love and care for each other because of our faith, and more. I was not raised in a very dualistic world as I was raised to see it as natural to talk to the universe through the constellations because God was listening.

It's through the lens that "science can only take you so far and then it's all mystery and God" that I take a quare[2] womanist look at the interplay of science and religion by centering the conversation on what makes us human and the way or ways culture and/or cultures figure large as we dare to challenge the boundaries between the academy and the community, between religion and the secular, and between theory and practice. I do so seeking to avoid what I sometimes find to be a retreat into abstractions that make it difficult for the discourse participants to articulate concrete concerns for justice. These abstractions are vexing *and* deadly because they invite/allow religion and science to erase, ignore, and devalue bodies of all sorts such as the human body I focus on in this essay. It is to, at times, fall into what Franz Fanon calls the "zone of nonbeing" and we see this in colored bodies, gendered bodies, sexual bodies, aged bodies, and more. This quare womanist look takes advantage of the ways in which queer theory gives me possibilities rather than a checklist of "dos and don'ts" in my work as a social ethicist. And while I am sympathetic to methodological precision in most cases, in this particular chapter, I invite you to join me in the shifting nature and sometimes messiness of queerness as a helpful conversation space to ruminate from a quare womanist perspective on science and religion in attempting to change the boxes that bind us. So I want to offer some ideas about ways to reorient the discourse so that we remain mindful of the living creation of which we are a part and embrace not only in our humanness but that of others as we sit in a whirling creation that is finite with glimpses of infinity and a God or sense of divinity who has and will outlast the boundaries of time.

LESS THAN HUMAN.

What makes us human has been a crucial question for Black folk in the western diaspora. This is the case because we began our lives as a people in the West being viewed as less than human or subhuman. This distortion was necessary to justify the modern enslavement of darker skinned peoples through the use of science and religion that assumed the dehumanizations of others to be an enlightened response to difference and a marker of progress—this was culture and ideology creating a troubling something in our common histories.

To live in a contested humanity is to constantly struggle with the question of what makes us human, what makes me, or you, a self, a person. As I was surfing the Internet in preparation for this chapter, I came across the following question posted on Yahoo! Answers three years ago:

> "Are white people and black people both human?"

My amusement at the question quickly turned to a "here we go again" moment as I read on:

> I was at the zoo the other day and there were a lot of black people. They were really loud and eating a lot of KFC and drinking a lot of grape soda and purple kool-aid. So when I got home I heard my parents talking about all of the monkeys from the zoo and that got me wondering. Are black people really just monkeys in disguise? Do they do stuff monkeys do like throw their poop? Did white people really evolve from these black people-monkeys? Why don't they

lock up these black people in the zoo if they are monkeys? I'm really confused right now. Please help, I need to know!

The first response to the question moved me from "here we go again" to anger:

> This is actually a very legit question . . . .
>
> Black people have different DNA then the rest of humanity, they are genetically built for warmer climate. There skin is darker for a reason, there skin produces more oils to block out bad rays from the sun, and to also soak up more heat, which in return makes them sweat more giving black people that "Zoo" like smell. They also lack melanoma, which is crucial to producing vitamin D. They have bigger noses and lips specifically made for warmer climate, they are use-to there own smells, black girls and black men find this "Zoo" smell attractive to one another.
>
> I am far from racist, the majority of my friends are black, but there is no reason to be "Politically Correct" here. Some people want to believe that everyone is made equal, this is what we are taught growing up, but the fact is every race is not equal, we are all made different. Some races have advantages over other races in certain area's, this is a fact. [3]

The exchange continues by citing the respondent's interpretation of the DNA studies of Dr. James Watson proving that Whites are more intelligent than Blacks and providing links that "prove" Blacks have different DNA from the Human Genome Project and concludes with the author's belief that the findings of the Human Genome Project will force society to reclassify "Negroes" as proto-humans and therefore proving that the Founding Fathers were correct to classify Blacks as 3/5th human.

This response, written by someone named NaZaReTh is stunning in its pseudoscience (and bad pseudoscience at that), is far too often compounded with suspect religious dogmas as it justifies cultural biases with "facts" that are nothing more than cultural biases that morph into racism and ethnocentrism.

It is crucial, I believe, to remember that the dialogues between science and religion take place within culture. Culture is a modern concept based on a term used by Cicero in classical antiquity—"*cultura animi*" (cultivation of the soul). This use of the term re-appeared in modern Europe in the 17th century referring to the betterment or refinement of individuals, especially through education. During the 18th and 19th century it came to refer more frequently to the common reference points of whole peoples, and discussion of the term was often connected to national aspirations or ideals. Some scientists such as British anthropologist Edward Tylor (1832–1917), a founding figure of social anthropology, used the term "culture" to refer to a universal human capacity.

An interesting aside about Tylor is that he saw religion as a survival or one of many customs a society retains as it evolves, that are unnecessary in the new culture as it emerges. In the 20th century, "culture" emerged as a central concept in anthropology that included human phenomena that cannot be directly attributed to our genetic composition. Today, we tend to think of culture in a variety of ways. In sociology, it is the total of the inherited ideas, beliefs, values, and knowledge, which constitute the shared bases of social action or the attitudes, feelings, values, and behavior that characterize and inform society as a whole or any social group within it: Black culture. In anthropology and ethnology, it is the total range of activities and ideas of a group of people with shared traditions, which are transmitted and reinforced by members of the group, for example the Mayan culture or a particular civilization at a particular period. In art, culture is the artistic and social pursuits, expression, and tastes valued by a society or class, as in the arts, manners, dress, etc. and/or the enlightenment or refinement resulting from these pursuits. And we go on to agriculture where culture is the cultivation of plants by scientific methods designed to improve stock or to produce new ones or the act or practice of tilling the soil. In microbiology, it is the experimental growth of microorganism or to grow microorganisms in a culture medium.

For my purposes, I refer to culture as socially transmitted behavior patterns, arts, beliefs, institutions, and other products of human thought and work. This contextualization of conversations about the interplay of science and religion is crucial as it helps ground us and encourages us to realize how important a conversation this is within the context of who we are and what makes us human and humane. For instance, like any social group in the United States and beyond, U.S. Blacks do meaning making and develop survival skills in the larger U.S. culture and the global arena. However, Black folk have lived and continue to live within a panopticon[4] in which people who often demonstrate meticulous illiteracy about Black lives watch them and judge them. Those who control the gaze of the panopticon refuse to acknowledge the impact of class, gender, and race on their observations and the ways in which illiteracy about Black lives quickly morphs into stereotypes that damn and eviscerate Black children, men, and women. It is a sad irony that often times those who control the panopticon also demonstrate meticulous ignorance about its existence and would protest (and believe) that they do not have the power or ability to exercise such control.

Historian of religion Paul Ricoeur is particularly helpful here as he encourages us to think larger than a singular culture but instead notes, "When we discover that there are several cultures instead of just one and consequently at the time when we acknowledge the end of a sort of cultural monopoly, be it illusory or real, we are threatened with destruction by our own discovery.

Suddenly it becomes possible that there are just *others*, that we ourselves are an 'other' among others."[5] This is an important reminder of the way in which culture works as there is not one acceptable culture be it high or low, but a plurality of them—each with its own audience. Each with socially transmitted behavior patterns, arts, belief, institutions. And this plurality is often mixed—not a plurality of separate cultures but a constant flux of cultures making and unmaking themselves through each other. So, I appeal to Ricoeur because we often carry out our dialogues about the interplay of science and religion as if culture is monolithic and far too often, that monolith is crafted from a White and Western and elite set of values, folkways, mores, and assumptions alone or as the norm.

But if we pause and consider the rich pluralism of our cultures, we realize that social groups complicate our dialogue of the science and religion interplay. For example, Black folks in the United States created and continue to nurture a culture against the backdrop of racial, social, sexual, and economic exploitation and injustice. Black folk then become agents who respond to their varying situations as well as affect these situations in crucial ways that are social, political, and theological. Although a part of the general American culture, Black culture has distinctive features that help maintain and shape the traditions of Black communities as they struggle to survive in the midst of the panopticon of structural and persistent oppression.

As such, common beliefs, value systems, and goals provide the framework for Black culture and provide the grist for that culture's participation in the dialogue between science and religion. This demands a conscious integration of authentic Black cultural realities in our theological musings to combat the gross stereotypes of Black lives. Also, a deeper, more careful methodology in the dialogue between science and religion not only combats gross stereotypes by being more attuned to cultural myopias enabled by the panopticon, but that doing so also changes the "lens" of scientific inquiry as well by exposing its meticulous illiteracies about what counts as scientific knowledge, or method, or evidence.

In sum, I am arguing that culture is a subjective something that informs how we understand science and religion. It shades us with the patina of our concrete materiality and the boundaries of our abilities to see others and the creation that surrounds us and holds us. It is important to remember how crucial a role the cultural ideologies of our times have on us such that it is wise to be circumspect about our ability to be objective across the science–religion continuum and beyond. And to consciously and rigorously integrate objectivity and subjectivity in our musings.

An abbreviated historical look at one trajectory of the interplay between science and religion that survives in the Yahoo! Answers question and response helps to guide us toward building a robust dialogue between science and religion as we explore what makes us human.

DISTORTIONS OF BLACKNESS THROUGH "SCIENCE"

The systematic assault on Black humanity is centuries old. It began, in part, with the interaction of the Portuguese with West Africa and the subsequent slave trade. This brought Europeans in frequent contact with dark Africans. The British could not understand the coloring of these folk because of the universal assumption that humanity stemmed from one source. This assumption was held in place by force of church and the interpretation of scripture.

Black skin became an indication of intellectual and moral inferiority. In the U.S. context, this form of hegemony began with the early settlers who frequently contrasted themselves with Blacks by using the term "Christian" more frequently than "English." Being English and being Christian became so intertwined that the Virginia Assembly of 1670 declared that "noe negroe or Indian though baptised and enjoyned their owne Freedome shall be capable of any such purchase of christians, but yet not debarred from buying any of their owne nation."[6] After 1680, a new term appeared—white.[7] To be Christian meant to be civilized, British, and white.

As historian Winthrop Jordan notes, The Virginia code of 1705 used Christian much more like a definition of race than of religion.

> And for a further christian care and usage of all christian servants, *Be it so enacted, by the authority aforesaid, and it is hereby enacted,* That no negroes, mulattos, or Indians, although christians, or Jews, Moors, Mahometans, or other infidels, shall, at any time, purchase any christian servant, nor any other, except of their own complexion, or such as are declared slaves by this act.[8]

Religious difference was *initially* more important than color, although colonists sought an explanation for the dark skin of the Africans. The answer they turned to most consistently was the sun. They assumed that the sun scorched the skin, was drawn to bile, or blackened the blood. This theory simply did not explain the fact that people living along the same longitude and latitude in the Americas did not have the same dark skin of equatorial inhabitants in Africa. Indeed, to follow this explanation to its logical conclusion meant that once Africans were in the northern countries they should start to lose their dark color. If it did not happen in the first generation of folks, it should appear in their offspring. By the mid-1600s, this explanation paled, as Blacks in Europe and the northern parts of the Americas did not whiten noticeably.

However, there were others who were certain that blackness was permanent and innate. For example, in 1646, the physician Sir Thomas Browne hypothesized that blackness was permanent and transmitted by the sperm although he was not sure why this was so. When the 1700s dawned, the mood was one of puzzlement and wild speculation posing as scientific theorems by mixing scripture, observation, and speculation. Interpretations of biblical stories fed into the subordination of blackness and Black people. Genesis 9:20–27 and Amos 9:7 became popular verses that "explained" Black inferiority.[9] As the century wore on, these explanations began to lose their popularity.

After the mid-1600s, some scientists abandoned the larger questions of color to explore the more manageable one of the physiology of skin color. Marcello Malpighi's pioneering work in microscopic anatomy (1648–1694) is a marker for the common view that black skin contained black fluid. The researchers were undaunted when they were unable to find such a substance.

Skin color became an independent rationale for enslavement. This may be because African assimilation, increased slave trade, and the arrival of larger numbers of Europeans outside England prompted the colonists to turn to physiognomic differences. All these theories and many subsequent hegemonic theories were based on the common assumption that the original color of humans was white. Therefore, blackness was degeneration from the original color. The idea of degeneration from primitive whiteness was seemingly confirmed by the fact that Black babies were born considerably lighter than their later skin color. Some, like Cornelis de Pauw, invented "facts" that they were not able to discover. In 1770, de Pauw announced that the Negro had dark brains, blood, and semen and the blending of dark and white semen caused the mulatto's intermediate color.

The 1758 edition of Carolus Linneaus' *Systema Naturae* was foundational in the natural sciences for many subsequent pseudoscientific explanations of race and ethnicity. Linneaus developed four categories for the species "Homo": African, American, Asiatic, and European. The African was described as black, phlegmatic, relaxed, hair black, frizzled, skin silky, nose flat, lips tumid, women's bosom a matter of modesty, breasts give milk abundantly, crafty, indolent, negligent, anoints himself with grease, governed by caprice. The American was described as red, choleric, erect, black hair straight and think, wide nostrils, freckled face, scanty beard, obstinate, content, free, paints himself with fine red lines, regulated by habit. The Asiatic was yellow, melancholy, rigid, black hair, dark eyes, severe, haughty, covetous, covered with loose garments, governed by opinions. The European was characterized as white, sanguine, brawny, hair abundantly flowing, blue eyes, gentle, acute, inventive, covered with close vestments, governed by customs.[10] It is indeed curious that the African was the only one in which women or their breasts are mentioned.

Johann Friedrich Blumenbach, an anthropologist, brought the aesthetic criteria and cultural ideals of Greece to the fore. Like Linneaus, he held that all humans belonged to the same species and races were varieties within that species. His 1775 work, *De Generis Humani Varietate Nativa* carried four species that moved to five in the 1781 edition, and the 1795 edition finally introduced precise language including the term "Caucasian."[11] In each edition, Blumenbach praised the symmetrical face as the most beautiful of human faces because it approximated the "divine" works of Greek art. He believed that this was yoked to climate—the more moderate the climate, the more beautiful the face. It is clear where this left Africans on Blumenbach's beauty scale.

Two popular and influential pseudosciences, phrenology and physiognomy evidenced the European biased character of scientific speculation. Phrenology, the reading of skulls had its most notable proponent in Pieter Camper, a Dutch anatomist, who discovered the facial angle in the 1770s. The facial angle permitted comparison of human heads by using cranial and facial measurements. The ideal facial angle was 100-degrees—achieved only by the ancient Greeks. Since a beautiful face, body, nature, character, and soul were inseparable, phrenology helped establish the "natural" superiority of the Europeans who had a 97-degree angle and the "natural" inferiority of Black people who were between 60–70-degrees—closer to the measurements of apes and dogs.

Later, in the early 1800s, Dr. Samuel Morton claimed that the human skull size determined the character of a person's intelligence *and* moral fiber.[12] Morton and other phrenologists claimed that because white people's skulls were larger than all others and Blacks had smaller skulls, whites were "naturally" superior to Blacks on intellectual and moral levels. Further, Morton claimed to provide documented proof that Blacks in ancient Egypt had been slaves. Therefore, Blacks had spent their entire existence on earth as slaves of the more intelligent races.

Johann Kasper Lavater, the father of physiognomy or the reading of faces, explicitly acknowledged that the art of painting was what informed his new discipline. Physiognomy equated visible physical characteristics with character and capacity. Lavater theorized that classical ideas of beauty, proportion, and moderation regulated the classification and ranking of human groups. He believed that Greek statues were models of beauty: blue eyes, horizontal forehead, bent back, round chin, short brown hair. Incidentally, these all parallel Camper's beautiful person.

Physiognomy was influential between scientists and artists. Goethe helped Lavater edit and publish his physiognomic formulations and Sir Walter Scott and others popularized them in novels. The striking thing about the method of this discipline is that it did not rely on detailed measurements, but the visual glance. Lavater was quite clear in his instructions, "Trust your first quick impression, for it is worth more than what is usually called observation."[13] He also developed an elaborate theory of noses that associated Roman and Greek noses with conquerors and persons of refinement and taste.

Major intellectual figures of the Enlightenment accepted these ideas without having to put forward their own arguments to justify them. In the Age of Criticism, this was quite odd for those who held reason and rationale on a sublime plane to accept uncritically the authority of naturalists, anthropologists, physiognomists, and phrenologists.

Montesquieu's remarks about Blacks suggest an ambiguous disposition toward some of these theories in his *Spirit of the Laws*. Although he opposed slavery, he turned to the sun theory to explain the existence of slavery because "there are countries where the excess of heat enervates the body, and renders men so slothful and dispirited that nothing but the fear of chastisement can oblige them to perform any laborious duty: slavery is there more recognizable to reason."[14] His conclusion resorted to the standard issue of the era: "It is impossible for us to suppose that these beings should be men; because if we supposed them to be men, one would begin to believe we ourselves were not Christians."[15]

The Scottish philosopher David Hume produced "Of National Characteristics" in 1748. This served as a major source of proslavery arguments and anti-Black educational propaganda. Hume was convinced that people living near the poles and in the tropical regions were inferior to those in temperate zones. He, like many before him and after him, appealed to the Greeks as the proving ground for his assertions. Greeks lived in a temperate climate; Greeks were the models of civilization, intellect, and morality. Therefore those who deviated from the ideal were inferior. In a footnote added in the 1753–54 edition, Hume stated:

> I am apt to suspect the negroes, and in general all other species of men (for there are four or five different kinds) to be naturally inferior to whites. There never was a civilized nation of any other complexion than white, nor even any individual eminent either in action or speculation. . . . .
>
> In Jamaica indeed they talk of one negro as a man of parts and learning; but 'tis likely he is admired for very slender accomplishments, like a parrot, who speaks a few words plainly.[16]

Immanuel Kant's ideas based his ideas heavily on Hume's claims. In his *Observations on the Feeling of the Beautiful and Sublime*, Kant names Hume directly:

> Mr. Hume challenges anyone to cite a simple example in which a negro has shown talents, and asserts that among the hundreds of thousands of blacks who are transported elsewhere from their countries, although many of them have even been set free, still not a single one was ever found who presented anything great in art or science or any other praiseworthy quality,. . . .So fundamental is this difference between the two races of man, and it appears to be as great in regard to mental capacities as in color.[17]

Even Thomas Jefferson did not escape this chorus of hegemonic supremacy masquerading as dispassionate reason. In *Notes on the State of Virginia* (1781–1782) Jefferson was clear:

> Comparing them by their faculties of memory, reason, and imagination, it appears to me, that in memory they are equal to whites; in reason, much inferior . . . and that in imagination they are dull, tasteless, and anomalous . . . Never yet could I find that a black had uttered a thought above the level of plain narration; never see even an elementary trait of painting or sculpture.[18]

There was a measurable lull in writing about Blacks between 1793 and 1807. However, the 1840 United States census was full of distortions about Black folk. This was the first census to record the number of mentally ill (classified as "insane and idiots"). The census suggested that freedom in the North led to widespread insanity and idiocy among northern Blacks. The ludicrous nature of such claims was highlighted by the fact that the census reported insane Blacks in counties where no Blacks lived. Dr. Edward Jarvis, a white doctor living in Massachusetts passionately rebutted the census findings in the *American Journal of Medical Science*. John Quincy Adams read Jarvis' findings and tried repeatedly to get the U.S. Congress to denounce the census but it never did. When Adams confronted John Calhoun about the errors in the census, Calhoun stated, "There were so many errors they balanced another and led to the same conclusion as if they were all correct."[19] Calhoun later used the 1840 census to bolster his argument to admit Texas to the Union as a slave state.

Several magazines and journals popularized these views. By the time of the Civil War, many in the United States public were convinced that Blacks were intellectually inferior because of subtle and blatant devaluations of the color black and the elevation of the color white. These cultural desires for superiority formed presuppositions prior to the science that not so subtly directed centuries of so-called scientific theories and findings regarding human difference. This case of scientific racism should stand as a warning to us every day, especially since, long after "science" has disavowed these claims, they live on and are communicated daily.

DEEPLY ENGRAINED CULTURAL VALUES AND ASSUMPTIONS.

Darwin changed the science of human difference by providing an explanation for the connection between species and their environments *and* between species themselves in external forces. Darwin concluded that forces outside the organism could change that species over time. Darwin's evolutionary theory did not feature fixed species created for given habitats by God, the march of progress, or some inner vital force. Instead, Darwin concluded that forces outside the organism could, over time, change species.

However what Darwin could not change is our struggle to transcend deeply engrained cultural values and assumptions that can morph into vexing racism and sexism and more. Darwin was a man situated in the complex worlds of his time and in the histories of science and pseudosciences. He had a deeply conflicted position on race. He argued on one hand the biological unity of all human groups in *The Descent of Man* and in other notes and publications while also unquestionably assuming the superiority of Europeans. Darwin predicted that intergroup competition would ultimately "At some future period, not very distant as measured by centuries, the civilized races of man will almost certainly exterminate and replace the savage races throughout the world."[20]

If I've learned anything from Darwin and those who have expanded and deepened his work or taken him down some dangerous rabbit holes, it is to be aware that we can fail at objectivity and devalue subjectivity prematurely in some spectacular ways as we seek to know ourselves and others if we do not remember that data can be wrong, values unexamined, cultural mores suspect, and religion devoid of moral backbone. Darwin and advocates of rival theories all concluded that there is a hierarchy of race based on mental and emotional competence. Darwin also espoused the general inferiority model of women's capabilities as he considered some of women's inferior traits to be fortuitous counterbalances to masculine characteristics.

To know ourselves as human asks each of us to practice our creative urges and methodological musings with a radical commitment, that is, a basic commitment to think through whether or not the interplay we seek in putting science and religion in conversation will lead us to a more just society. The work we do and the conversations we have are embodiments of our values, hopes, compassion, interconnectedness, rituals—our isness. Knowing the worrisome parts of our history should not deter us from engaging in robust conversations. No, they give us guidelines and warning signs and symbols for what to look out for to keep us pointed in directions that lead people to an expansive sense of self that is not the narcissistic self of Aaron James' *Assholes: A Theory.*[21] Instead, these selves coming to wholeness form divine/human communities based on love and hope and are pointed toward justice. The interplay of religion and science can open us up to new conversations that can get us there. And this is where I believe queer theory, with its disruptive *jouissance* helps us to disrupt hegemonic discourses in the sciences and in religion by methodically destabilizing antagonistic binaries such as heterosexual/homosexual, gay/lesbian, feminine/masculine, colored/white, poor/wealth, profane/sacred with a quaring eye that disrupts the glare and judgment of the panopticon. By employing a quaring eye in this conversation, I want queerness to stay imprecise, to engage in theoretical rope-a-dope; to challenge us to think through *why* we need to spend time unsettling science and religion through the lenses of queer studies. It is precisely because quareness unsettles categories of gender, sexuality, race, ethnicity, class, and so on as it also challenges notions of empirical investigation, research and theory, direct investigation, coherent chains of rigorous reason, replication, transparency, and scholarly debate that it does not allow any of us to sit too comfortably in the conversations that I trust will unfold.

What quaring does in my work (and I hope as I have deployed it throughout this chapter) is to provide the background noise to help me point to the inequalities, the glitches, the imprecisions, the debates, and more. It is an important part of the womanist methodology I employ by using a conscious matrix of class, gender, sexuality, sexual orientation, race/ethnicity to unseat and disrupt if not help transform injustices into justice-making scholarship and living.

I believe that when we engage in these new conversations we will expand our understanding of the demands of public life and how we can and must pursue just relationships within the context of creation. This urges us to move beyond our various social locations so that what is seen and known and assumed when we use alleged universal categories or objective measures may, in fact, be judging others (and ourselves) against the yard stick of mini-mes that pose as intellectual rigor but are nothing short of ideological blasphemies that are tone deaf to differences, declare diversities as suspect, and croon off key lullabies about post-racial America—a completely imaginary and fictional notion that belongs with unicorns, pixie-dust, and separate but equal.

To know ourselves as human and to move to a more constructive, holistic notion of self and personhood places theological anthropology squarely within the conversations that we need to engage in. Quareness opens a way for us to craft vibrant conversations about the interplay between science and religion that is not terrified of

the curve of our hips
the arch of our backs
the slow swing in our walks
the glide of our fingers
the fire in our eyes
the coil of our hair
the deep moans and shouts of our ecstasies
the bottomless welling cries of our sorrows
the slow bend of our smiles
the precision of our minds
the sass of our talk

Not terrified of our bodies that carry our past, our present, our future—perfectly and imperfectly.

To do so means that we intricately factor in the impact of culture on science and religion as we strive to build on these to create rational narratives that embrace poetry and beauty that engage the heuristics of religion through critical thinking about the issues of our day such as global warning, evolution in public schools, stem cell research, human cloning, and more. We are dangerous when we believe that it is immoral or unfaithful or just plain wrong to engage science from our various lenses of religion. And we are dangerous when we believe that it is unscientific or irrational or just plain wrong to engage the horizons of awareness, possibility, and desire that our various lenses of religion bring to science.

If we can quare it, we just might be able to change it and get ourselves out of some of the stultifying boxes we find ourselves in and place others into. Queer is an umbrella. It is fluid, it is persistent in challenging strict codification, if not absolutely refusing to be hemmed in and penned up. And because it is this kind of "trouble" we need to be aware that quaring is not a side door into some form of justice for all. Quare is messy and just as vulnerable to the ways that culture pushes us into being some of our worst selves. So, I am reminded as I close to urge us to also beware and alert as we try to learn from queer studies.

Science is not and cannot be a source of ultimate moral authority. Reason untethered from the demands of diversity and an expansive understanding of whom and what makes up the social order is not a foundation for expansive moral reasoning. Religion without a sense of our own situatedness stifles the moral imagination so that we are unable to be mindful of the living creation of which we are a part and embrace not only our humanness but that of others as we sit in a whirling universe that is infinite, I suspect, and a God who has and will outlast the boundaries of time. Some like to put God in the gaps of what science cannot know, or cannot yet know, as an inferior placeholder for future knowledge. But some understand that this dualistic approach to science and religion misses the deeper interaction of situatedness, moral imagination, and careful observation. Scientific exploration can only take us so far and then it's all mystery and God.

WORKS CITED

Adams, Charles Francis. ed., *Memoirs of John Quincy Adams, Comprising Portions of His Diary From 1795–1848, Vol 12*. Philadelphia: J.B. Lippincott & Co., 1874–77.

Bentham, Jeremy. *Panopticon (Preface)*. Edited by Miran Bozovic. *The Panopticon Writings*. London: Verso, 1995.

Darwin, Charles. *The Descent of Man and Selection in Relation to Sex*, Vol. 1. London: John Murray,1871.

Hume, David. *Essays: Moral, Political, and Literary, Vol. 1*. Edited by T. H. Green and T. H. Grose. London, 1875.

James, Aaron. *Assholes: A Theory*. New York: Doubleday, 2012.

Johnson, E. Patrick. "'Quare' Studies, or (Almost) Everything I Know About Queer Studies I Learned from My Grandmother." *Text and Performance Quarterly* 21, no.1 (January 2001): 1–25.

Jordan, Winthrop D. *White Over Black: American Attitudes Toward the Negro, 1550–1812*. Baltimore: Penguin Books, 1968.

Mosse, George L. *Toward the Final Solution: A History of European Racism*. New York: Howard Fertig, 1978.

Nugent, Thomas, trans. *The Spirit of the Laws by Baron de Montesquieu, Vol. 1*. New York: D. Appleton and Company, 1900.

Ricoeur, Paul. *History and Truth*. Evanston, IL: Northwestern University Press, 1965.

Said, Edward. *Orientalism*. New York: Vintage Books/Random House, 1994.

Stanton, William. *The Leopard's Spots: Scientific Attitudes Toward Race in America 1815–1859*. Chicago: University of Chicago Press, 1960.

Townes, Mary M. and Dugald E. S. Brown. "The Involvement of pH, Adenosine Triphosphate, Calcium, and Magnesium in the Contraction of the Glycerinated Stalks of Vorticella." *Journal of Cellular and Comparative Physiology* 65, no. 2 (April 1965): 261–269.

West, Cornel. *Prophesy Deliverance!: An Afro-American Revolutionary Christianity*. Philadelphia: The Westminster Press, 1982.

Williams, Delores S. *Sisters in the Wilderness: The Challenge of Womanist God-Talk*. Maryknoll, NY: Orbis Books, 1993.

NOTES

1. Mary M. Townes and Dugald E. S. Brown, "The Involvement of pH, Adenosine Triphosphate, Calcium, and Magnesium in the Contraction of the Glycerinated Stalks

of Vorticella" *Journal of Cellular and Comparative Physiology* 65, no. 2 (April 1965): 261–269.

2. Quare is the southern equivalent of queer. It is usually derogatory unless one is actually from the south and speaking about a homosexual. Pronounced: Qw-AR. See also E. Patrick Johnson, "'Quare' Studies, or (Almost) Everything I Know About Queer Studies I Learned from My Grandmother," *Text and Performance Quarterly* 21, no.1 (January 2001): 1–25.

3. https://answers.yahoo.com/question/index?qid=20120311192610AARaQ7h

4. English philosopher and social theorist Jeremy Bentham designed the panopticon in 1785. It is a type of prison building designed to allow an observer to observe all prisoners without the incarcerated being able to tell whether they are being watched. See Jeremy Bentham, *Panopticon (Preface)* in *The Panopticon Writings*, ed. Miran Bozovic (London: Verso, 1995), 29–95.

5. Paul Ricoeur, *History and Truth* (Evanston, IL: Northwestern University Press, 1965), 278.

6. Winthrop D. Jordan, *White Over Black: American Attitudes Toward the Negro, 1550–1812* (Baltimore: Penguin Books, 1968), 94.

7. Ibid., 95.

8. Ibid., 94.

9. In the Genesis passage of the Noah and Ham story, blackness is never mentioned. However the colonists interpreted the story so that blackness and slavery were equated. Jonathan Edwards, Thomas Hooker, and George Whitefield were among the many preachers who associated the devil's habits with the color black

10. Edward Said, *Orientalism* (New York: Vintage Books/Random House, 1994), 119.

11. Cornel West, *Prophesy Deliverance!: An Afro-American Revolutionary Christianity* (Philadelphia: The Westminster Press, 1982), 57.

12. See Delores S. Williams' discussion of Morton in *Sisters in the Wilderness: The Challenge of Womanist God-Talk*, (Maryknoll, NY: Orbis Books, 1993), 95–96.

13. Quoted in West, 59. See also George L. Mosse, *Toward the Final Solution: A History of European Racism* (New York: Howard Fertig, 1978), 25.

14. Thomas Nugent, trans., *The Spirit of the Laws by Baron de Montesquieu*, Vol. 1 (New York: D. Appleton and Company, 1900), 240. See also Jordan, 261.

15. Quoted in West, 61.

16. David Hume, *Essays: Moral, Political, and Literary*, Vol. 1, eds. T. H. Green and T. H. Grose (London, 1875), 252.

17. Quoted in West, 62.

18. Quoted in West, 62.

19. Charles Francis Adams, ed., *Memoirs of John Quincy Adams, Comprising Portions of His Diary From 1795–1848,* Vol 12 (Philadelphia: J.B. Lippincott & Co., 1874–77), 61–62. Also quoted in Williams, 95. See also William Stanton, *The Leopard's Spots: Scientific Attitudes Toward Race in America 1815–1859* (Chicago: University of Chicago Press, 1960). This is a study of the ways the federal

government has used dishonest strategies to perpetuate the notion that Black folk are the intellectual midgets to Whites.

20. Charles Darwin, *The Descent of Man*, 1871, Vol. 1, 201.
21. Aaron James, *Assholes: A Theory* (New York: Doubleday, 2012).

Chapter Three

Thinking through Three Revolutions

Religion, Science, and Colonialism

Lisa Stenmark

From the late-15th to the mid-16th centuries, Europe experienced three revolutions, each of which indelibly shaped Europeans' perceptions of the world, as well as their interactions with other people and cultures. The geographical revolution began in 1492 when Columbus "discovered" the Americas and continued through the 18th century and the Age of Discovery. This revolution expanded Europeans' understanding of the size, shape, and geography of the world, and brought them into contact with radically different peoples and cultures. The second revolution was religious, the Protestant Reformation, which (for argument's sake) began in 1517 when Martin Luther nailed The 95 Theses to the door of the Castle Church in Wittenberg. This revolution expanded Europeans' understanding of religious possibilities and undermined religious authority. Finally, the Scientific Revolution, beginning (again, for argument's sake) in 1543 with the publication of Copernicus' *De Revolutionibus Orbium Coelestium,* expanded Europeans' understanding of the nature and purpose of the natural world and changed the way they interacted with it. These three revolutions so profoundly changed the way Europeans thought about the world, their relationship to it, and to other peoples, that in a period of one hundred years, what an educated European "knew" about anything and everything had completely changed.

These revolutions were not distinct, but were instead intertwined with each other (and with other revolutions), with each one shaped by and shaping the others. Within the religion and science discourse, we focus on the relationship between two of these revolutions—the Protestant Reformation and the Scientific Revolution—but ignore the global revolution and colonialism. This is part of a larger tendency to take a narrow view of "religion" and "science," treating them as discrete practices existing in isolation from broader cultural and historical developments, which provides an impoverished view of religion and science and the relationship between them. In this narrow view, if the relationship between the three revolutions—and thus between religion, science, and colonialism—is considered at all, it is seen as nothing more than an accident of history. This narrow view exaggerates conflicts between Western religion and Western science, obscuring their complicity in the colonial enterprise and the ways that Western science develops by distinguishing itself not *merely* from Western religion (creating a binary of science and religion), but also from many ways of knowing, creating a series of overlapping binaries of gender, race, and class. These binaries are not external to "science" but rather define it. By paying attention to the relationship between religion, science, and colonialism, it becomes clear that particular ways of understanding the relationship between religion and science have been supported by and continue to support colonial ideologies and thus undermine our efforts at justice and human (and planetary) flourishing.

In this chapter, I begin by describing the relationship between Western science and colonialism, a relationship that is practical and institutional, but also ideological, as the successes of Western science (and technology) were key justifications for colonialism. I will then argue that these scientific justifications exist within a structure that duplicates the structure of Western religions—what Martin Jaffe calls the "structure of elective monotheism."[1] Jaffe's insight is that the key element in Western religion is not merely the existence of a singular, universal deity (monotheism), but the existence of a community which has exclusive possession of a singular, universal truth, and which is obligated to convince others to accept that truth and act in accordance with it. Elective monotheism supported colonial ideologies—first through claims of a unique religious exclusivity, which were replaced by claims of a unique *scientific* exclusivity. But, whether religious or scientific, what Jaffe calls the *structure* of elective monotheism remains unchanged, and it leaves few options for addressing divergence (from that truth) or difference, other than subjugation or eradication. Moreover, this structure remains part of the discourses of religion and science and, thus, of the Science and Religion Discourse (SRD). I will conclude by exploring resources within queer theory (QT) and Postcolonial Literature for disrupting this structure and opening up new possibilities for our encounters with divergence and difference. The concept of hybridity/the hybrid—which Jessica Langer describes as the "hopeful monster"—provides an alternative to the destruction of either the colonized or the colonizer, challenging the structure of elective monotheism and offering a more inclusive approach to understanding the relationships between religion and science.

First, I want to clarify some terminology. I will use "science" broadly to refer to activities and practices aimed at producing knowledge of the natural world, while "Western science" describes a *particular* set of practices. I don't mean to imply that "Western science" is monolithic, nor do I want to center the practices of Western science in relation to other "sciences." There are many sciences, and many sciences produced in "the West," but the term "Western science" describes a set of identifiable practices which characterize a particular worldview and a particular way of producing knowledge, including such things as systematic experimentation, the "scientific method," a search for underlying principles, and ideas about objectivity.

I will also address underlying ideas, assumptions, and attitudes associated with European colonialism. Implicit in all our practices and institutions are beliefs that are largely unacknowledged, and colonialism is no exception. As Edward Said argues, both imperialism (the domination of distant territories) and colonialism (establishing settlements on those territories) "are supported and perhaps even compelled by impressive ideological formations that include notions that certain territories and people *require* and beseech domination."[2] There are a number of ways to describe the ideological formations of colonialism—imperial imaginaries,[3] ideological phantasies[4]—but in less academic terms, one might ask: what underlying beliefs would lead people to think it is a good thing—both practically and morally—to travel half way around the world, subjugate the people there, exploit their resources, and impose political, economic, and religious systems on them against their will? I will refer to this set of underlying beliefs as an "imperial ideology." Not all imperial ideologies are the same—there have been different imperial powers, and thus different imperial ideologies—but in this chapter I will use the term to refer to a particular set of beliefs associated with Western imperialism and colonialism. I know this term will be off-putting for some: we do not like to think of ourselves as driven by imperialism or ideology. But it is necessary to talk about this honestly because, as I shall argue, it is clear that Western science, along with Western religion, has been and remains complicit in maintaining imperial ideologies. Until we can name it, we can't change it, and we will continue to repeat the destructive and dehumanizing attitudes and actions of the past without understanding *why*.

SCIENCE AND COLONIALISM

In the past twenty years, scholarship exploring the relationship between Western science and Western colonialism has demonstrated that the development of each was integral to the development of the other—Western science and the scientific enterprise were not incidental to colonialism, nor were they "cultural afterthoughts." They were instead "part and parcel with colonialism from the beginning."[5] Science and technology directly supported the colonial enterprise through such things as advances in ship design and technology—"an area of technology in which the Europeans had few rivals by the fifteenth century and in which they reigned supreme by the seventeenth"[6]—along with the knowledge of winds, tides, and navigation necessary for exploration, trade, and colonization. Moreover, once Europeans arrived in the "New World" science provided the knowledge necessary for survival and entire disciplines, such as botany and (tropical) medicine, developed in response to the needs of empire.

Colonialism also created and shaped Western science. Colonial interests poured significant funds into research that would help establish and maintain colonies, control native populations,[7] and generate profits, for example, by identifying plants and animals amenable to large scale production, including cochineal and quinine. Colonialism also supported Western science by providing access to a vast array of data made available by colonial expansion. This included medical data from diverse populations, measurements of the speed of pendulums at different locations, data to resolve the problem of longitude, and "big science" projects such as the global coordination of measurements of the transit of Venus which established the distance between the earth and the sun and thus determined "the real dimensions of the solar system for the first time with any accuracy."[8] Western scientists also had access to data produced by indigenous scientists—although local science was devalued as "folk" tradition, superstition, or mere technical knowledge, European scientists still appropriated that knowledge for their own projects. At the same time, indigenous knowledge that was *not* deemed useful to colonial interests was systematically destroyed through lack of funding, exclusion from educational curriculum, and even criminalization, eliminating alternative explanations and approaches.[9] The simultaneous appropriation and repression of indigenous science benefitted Western science to the detriment of indigenous science as data and resources flowed from colonies to Europe, while the flow was impeded within and between the colonies themselves. This had consequences in terms of resources, but also increased the status of Western science to the detriment of indigenous sciences, contributing to the perception of the uniqueness of Western approaches. In the process, Western science became, simply, science.

The relationship between Western science and colonialism existed on an ideological level as well. In *Machines as the Measure of Man*, Adas observes that by the 19th century, Europeans' perceptions of "the material superiority of their own cultures, particularly as manifested in scientific thought and technological innovation" dominated European thought and "shaped their attitudes towards and interactions with peoples they encountered."[10] This sense of superiority was one of the "key components of the civilizing-mission ideology that both justified Europe's global hegemony and vitally influenced the ways in which European power was exercised."[11] This sense of superiority was not new, but prior to the 19th century Europeans "viewed their Christian faith, rather than their mastery of the natural world, as the key source of their distinctiveness from and superiority to non-Western peoples,"[12] and it was the evangelizing impulse within Christianity that motivated and justified early imperialism. By the end of the 19th century, however, Europeans had come to believe that the "unprecedented control over nature made possible by Western science and technology proved that European modes of thought and social organization corresponded much more closely to the underlying realities of the universe then did those of any other people or society, past or present."[13] Science developed *only* in Europe, because it was *only* in Europe that science and reason had triumphed "over the forces of superstition and ignorance"; while other cultures remained "unable to extract their scientific and technological concerns" from superstition and religion, lacking the "political climate for self-critical thought."[14] Control of the environment became the line between savagery and civilization, and the evangelizing impulse was now directed towards "civilizing" indigenous peoples and improving the conditions of their lives by, first, controlling nature (and them) and later, perhaps, teaching indigenous peoples how to control nature themselves (there was disagreement as to whether "savages" would *ever* be able to master these processes). Science did not merely provide tools for colonialism, it legitimated colonialism by providing the justification for devaluing and delegitimizing other cultures, and their knowledge of the world. Thus, the mutual support of Western science and colonialism extends to the ideological level: colonial expansion contributed to the growth and success of Western science, while the growth and success of Western science justified colonialism.

FROM ELECTIVE MONOTHEISM TO THE CIVILIZING MISSION

So far I have focused on the connection between Western science and colonialism, but Western religion was also part of the "ideological formation" of colonialism. This included an institutional connection, for example, missionary priests were often the first scientists, and religious schools included Western science in their curriculum.[15] But there was a deeper connection because as the perception of European superiority shifted from Christianity to Western science and technology, the evangelizing impulse to convert native populations never went away: religious and scientific justifications went hand in hand. Missionaries saw superiority of Western science as a justification for mission work, and used those arguments to win converts and convince local populations to send their children to mission schools.[16] The connection is even deeper, however, because scientific justifications did not merely replace religious justifications, they became part of a larger framework that remained, essentially, religious. Scientific justifications, like religious justifications, were "grounded in the presuppositions that there are transcendent truths and an underlying physical reality which exists independent of humans, and both are equally valid for all peoples."[17] For Christianity, of course, that transcendent truth was a single, universal God and a single, universal religion, available exclusively to [European] Christians. In Western science, it was a belief in a single, universal nature and a single universal science, available exclusively to Europeans. The thread that runs through all this is a belief in a single, universally valid truth; a single, universally valid knowledge of it; and the existence of a select group with access to that knowledge. Thus, while the content of the justification changes, the structure does not, and it is a structure that justifies domination and imperialism.

These justifications reflect the symbolic structure of Western thought which Martin Jaffe describes as "elective monotheism." Jaffe argues that the key element of Western religion is not merely the uniqueness and unity of the divine Being, but the exclusiveness of the communication and the community that responds to that communication. While the specific elements of this structure differ within and between the various Western traditions, according to Jaffe the structure of elective monotheism remains the same, with two dimensions (see Figure 3.1): a vertical dimension in which a single, universal Being discloses "a total truth about God and the world" to a specific individual or community, and a second dimension in which the community, as a result of being the exclusive recipient of this unique and universal communication, becomes the center of God's action in the world. The community is called to bridge the gap between the world and the Being, transforming the human order by communicating this unique and universal truth to the entire world which is eventually absorbed into this community. There will be resistance, both internally and externally, but the community must persevere against those who have not received or refuse to accept the unique and universal message. Thus the "evangelizing impulse" is built into the ideological structure of elective monotheism, and in each tradition it is tied to territorial expansion as "political dominance" becomes one way of spreading the truth and expanding the community.[18] The uniqueness and universality of the communication and, thus, of the community entrusted with it—*elective* monotheism—suggests that while there is universal truth, it is not universally

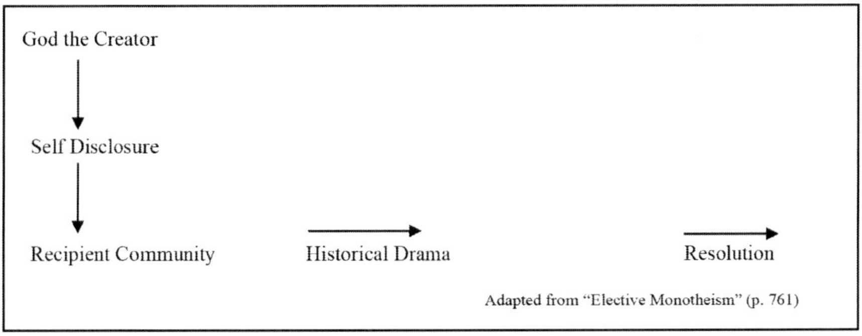

Figure 3.1.
Graphic courtesy of the author.

available. Those who disagree with the selected community are in the wrong, and the only way to deal with this divergence is through eradication or subjugation. No wonder Jaffe calls this "a remarkably fertile symbolic universe for legitimating violence."[19]

This pattern is repeated in Western science, where nature is singular and reveals her secrets exclusively through the techniques and practices of the scientific method, making the scientific community the exclusive possessor of knowledge necessary to understand and control nature. As with Western religion, knowledge of the world is universal but not universally available; it is only available through Western science. Thus, scientists have an obligation to manifest this knowledge in the world, to spread its benefits. The evangelizing mission became the "civilizing mission" of the French and the "white man's burden" of the British, which were both supported by the belief that Europeans alone had access to knowledge which no one else possessed, which gave them both the right and the responsibility to occupy the earth. "One God, One Revelation, One Religion" becomes "One Nature, One Truth (about it), and (thus) One Science," "and such a science can in principle reveal the complete, unitary and harmoniously integrated truth about a reality that is ordered in such a way as to be available for such an accounting."[20]

Jaffe's description of the symbolic structure of elective monotheism provides insight onto the occasional conflict between Western science and religion, but it also provides insight into the connection between religious and scientific justifications for colonialism, including the importance of territorial expansion under the umbrella of spreading knowledge. Within the structure of elective monotheism, Western science is not only superior, because it is unique, universal, and exclusive, that superiority justifies imposing it on others, in spite of resistance. This structure gives meaning to the ideologies of colonialism, including the narrative of progress, in which European

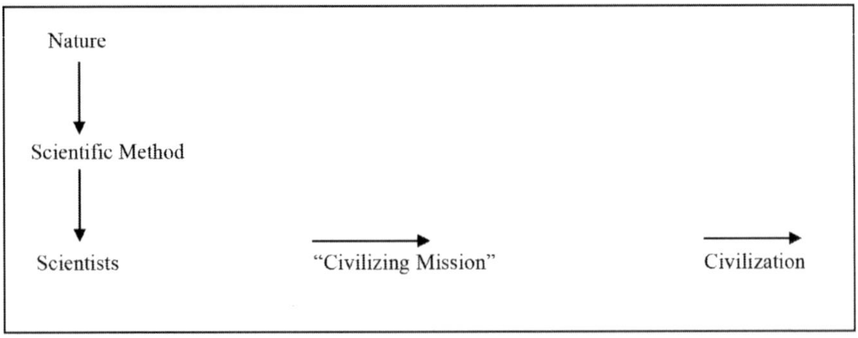

Figure 3.2.
Graphic courtesy of the author.

knowledge and culture is the locus and impetus for history. It gives rise to a kind of binary thinking embedded within colonialism, a center-periphery model in which all the action and agency is in the colonizing, metropolitan center which represents order, science, and modernity while the periphery represents chaos, superstition, and backward thinking. "Christendom" is merely replaced by a rational scientific culture.

These insights are not particularly new, yet they rarely if ever make it into academic discussions of the relationship between religion and science. We focus on the (occasional) hostility between Western science and religion—the relationship between *two* revolutions—but ignore the third. This causes us to overemphasize differences between Western sciences and religion, ignoring the ways they worked together to support colonial ideologies. Consideration of the third revolution realigns our thinking, because from the outside, Western religion and science are not in conflict but are instead complicit in delegitimizing and silencing indigenous traditions. But despite the direct connection between the advancement of Western science and the destruction of non-Western religious traditions we rarely focus on this hostility. Indigenous traditions are treated (explicitly or implicitly) as less legitimate—often because they are "local" as opposed to universal—than Western traditions, religious or scientific, which are treated as uniquely universal. In addition, the third revolution reminds us that at the same time science is defining itself over and against religion, it is also defining itself over and against indigenous practices. Thus, there is a series of binaries in which science—and the Western, rational, scientific, and civilized male—is at the center and everyone else is in the periphery. These interrelated binaries of race, gender, class (and more) are not incidental to science—they are an inherent part of our understanding of what science is and isn't.

This matters because colonial ideologies are not ancient history, they are not even history: the belief in a transcendent truth which can direct our actions persists, as does the belief that it is singular and universal, and that getting others to accept and act in accordance with that truth is "helping." The problem is not that we believe this consciously, but that these are embedded in our institutions and practices, exerting an influence on us, culturally and individually, creating relationships of domination and superiority in which we implicitly and explicitly devalue other knowledge systems and justify exploitation in the name of "development." We will not be able to act differently until we acknowledge, dismantle, and deconstruct these ideologies within the practices and assumptions of our various discourses, and thus within the science and religion discourse.

I am not suggesting that we cast aside Western science, or that it needs to be replaced by another system of knowledge—this is not an either/or choice—I am instead advocating for the inclusion of multiple knowledge systems instead of measuring all knowledge by the standards of Western science. The primary arguments in defense of Western science often come down to the assertion that there is a neutral core or content of science that is free from cultural bias. That core may indeed exist, but it is irrelevant, because to think it can exist *for us* outside of the structure and organization of science—or of the broader culture—is disingenuous. We cannot separate "Western science" from the "history of Western science," any more than we can distinguish "Christianity" from the history of Christianity, because the institutions, practices, and theories of knowledge produced by history are always still in place. The question is not whether and how accurate the knowledge produced by Western science is, the question is whether and how the conceptual and institutional structures of our various disciplines and discourses reinforce ideas and practices that undermine our efforts at social, economic, and environmental justice. To be more blunt, my concern here is not epistemology, it is justice.

The historical, institutional, and ideological connection between these three revolutions suggests that the task of rethinking the relationship between religion and science cannot ignore the history of colonialism. Moreover, because colonial ideologies are so intertwined with the ways we think about religion and science and the relationship between them we will not only need to expose, examine, and deconstruct these modes of thought, we will need to construct alternatives. This involves a kind of apocalyptic thinking in which we critically examine our current position and also attempt to embody new approaches and understandings. One part of this will have to be challenging universal and transcendent conceptions of truth claims. Challenging those certainties is one area where queer theory will be especially helpful. This is not a prelude to replacing it with another system, because this would merely shift the symbolic system around within the existing structure of elective monotheism. Thus, in addition to critically addressing the universality of truth claims, we also need to address the exclusivity of the community in relation to those truth claims, because, combined, these make it impossible for us to deal with difference: we can eradicate it through negation, dismissal, or appropriation, or we can subjugate it by placing it within a system in which some knowledge claims are better than others (or, using scientific language to explain what they "really" mean). In the final section of this chapter, I will explore one way to decenter Western science that also creates opportunities for more productive engagements with multiple knowledge systems, the concept of hybridity.

HYBRIDITY AND THE HYBRID

Hybridity and the hybrid are often associated with postcolonial literature, where it reveals the multiplicity of colonial identities and colonial cultures, challenging essentialist notions that led to the destruction and subjugation of indigenous peoples and their cultures. Along with queer theories, hybridity in postcolonial literature challenges conceptions of identity as fixed and congruous, revealing the ways it is a fluid and multi-faceted social construct. These challenges have been extended to include all manner of fixed certainties, suggesting that what we perceive as "just the way things are" is actually "just the way we have made things." The distinction between postcolonial and queer theories is largely a matter of focus and emphasis, with queer theory starting from the erasures caused by fixed categories of gender and sexuality, while postcolonial literature starts from the erasures caused by the forced imposition of an alien (colonial) culture. Inevitably, each trajectory has blind spots, so any effort to critically examine our understanding of the relationship between religion and science will need to draw on a number of critical discourses.[21]

The contributions of hybridity, its connection to queer theory, and the value of using multiple trajectories to rethink our approach to the relationships of religion and science are all suggested by the work of Gloria Anzaldúa and others who have located themselves at the intersections of race, gender, sexuality, and class, in order to challenge claims of primacy from any particular location. In *Borderlands*, for example, Anzaldúa explicitly connects the experience of being *mestiza*—a "hybrid progeny of mixed race"—with being queer: "as a *mestiza* I have no country, my homeland cast me out . . . as a lesbian I have no race . . . but I am all races because there is the queer of me in all races."[22] Like the mestiza, queers exist "on both sides at once," neither here nor there, but simultaneously nowhere and everywhere.[23] The *mestiza*/queer receives "multiple, often opposing messages," facing the dilemma of the mixed breed: which collectivity does she listen to? The result is a struggle of borders, "*un choque*, a cultural collision" that produces a state of "mental *nepantalism*—torn between ways."[24] In order to heal the conflict, she must learn to be "on both shores at once." This begins with "a conscious rupture" with what is oppressive in "all cultures and religions." She must then communicate that rupture, document the struggle, and reinterpret history, using "new symbols" and "new myths." In the process, she "strengthens her tolerance (and intolerance) for ambiguity. . . . She surrenders all notions of safety, of the familiar. She becomes a *nahual*, able to transform herself into a tree, a coyote, into another person. She learns to transform the small 'I' into the total Self."[25] This is the "new Mestiza consciousness," which is "the consciousness of the Borderlands,"[26] a source of pain, but also the source of a continuous creative motion that breaks each new paradigm, a consciousness that "in our best hopes, bring us to the end of rape, of violence, of war."[27]

There are several aspects of Anzaldúa's discussion of the borderlands consciousness that I want to highlight. First, her understanding of identity and community is explicitly an alternative to Western notions of purity and duality, in which unity and coherence is achieved by similarity of thought, which requires exclusion. Anzaldúa does not want us to define ourselves by who and what we exclude, but by who and what we include. As a result, we no longer need to gloss over or eliminate differences, because differences do not threaten our sense of unity or coherence, they expand them. The second point is that by including these multiple locations, it becomes possible to develop a kind of critical consciousness, one that can evaluate many positions. Third, this is not a settled position. Quite the opposite, as Anzaldúa is clear that the borderlands include instability, contradiction, and pain. But this struggle also opens a space for new and creative possibilities—there is a tension that is both destructive and creative; the borderlands create opportunities to offer up new possibilities for what can be. Fourth, the *mestiza* consciousness is not inherent in persons who are "mixed race," or queer, it is the experience of being outside of a fixed location that makes the development of this consciousness *possible*, and perhaps necessary for survival, but not inevitable because the mestiza consciousness developed only when she *embraces* the contradiction and struggle of this location. Extending Anzaldúa's argument, we can see difference itself as a kind of ontology—we are all different, regardless of gender, race, or sexuality. Some differences are hidden in the center, while others are highlighted at the margins. Those in the margins live in the borderlands already, but those of us in the center can intentionally locate ourselves in the borderlands by learning to embrace difference and contradiction. I don't want to glibly suggest that this is an easy process, but the broader point is that we are not defined by our difference. We are defined by how we choose to respond to difference: we can reject it or embrace it.

Many of these same themes are present in queer of color critiques and global LGBTQ communities, each of which locate themselves at the intersection of race, gender, and sexuality (and often class), simultaneously occupying multiple locations as a way to disidentify with—and thus deconstruct and transform—categories of gender, sexuality, and race that are too rigid.[28] LGBTQ communities outside of the West, for example, who occupy the "(queer) hybrid spaces between national borders" are exploring possibilities for expressing gender and sexual identities outside the confines of Western categories. "Global queering" neither mimics Western culture, nor does it return to "traditional" ideas, but instead blends both Western and traditional concepts and practices regarding sex, gender, and race in ways that create new possibilities for expression. The "tension between globalized and localized homosexualities . . . help destabilize related fixed social epistemes pertaining to race, nationality, gender and social class *in addition to* sexuality."[29] Here again, it is by exploring internal contradictions that hybridity challenges each new paradigm that emerges. Hybridity queers the queer.

Hybridity is an important aspect of postcolonial literature, which is itself an example of hybridity, as authors within former colonies attempt to express, transform, and create a postcolonial culture using a language—and a form—that is itself a legacy of colonialism. Hybridity is also important to speculative fiction (SF), and the hybrid has been an important figure in SF since the beginning.[30] Thus, it should not be surprising that the hybrid is important in postcolonial speculative fiction (PCSF)[31]—a subgenre of postcolonial literature and SF, a hybrid within a hybrid in which indigenous authors have located themselves in a genre devoted almost exclusively to exploring the significance of the Western scientific consciousness, but do so in a way that simultaneously challenges its boundaries. PCSF "fuses indigenous sciences with the latest scientific theories,"[32] blurring the lines between "real" and "unreal," animal, spirit, human, and machine in ways that challenge Western perceptions. It is this very hybridity that makes PCSF particularly relevant for thinking about the relationship between religion, science, and (post) colonialism, because it draws these elements together into a single narrative, drawing the discordant paradigms of Western science and indigenous scientific literacies into a coherent whole that does not eliminate or subordinate either.[33] Postcolonial literature focuses on difference in order to distance colonial cultures from the cultures of the colonized.

The colonial narrative insists that cultural borrowings are only one way—the West bestows it cultural gifts upon other, less advanced, peoples, while it is not (or *should* not be) changed by them (because this would diminish Western culture). The result is a center/periphery model in which the culturally, technologically, and scientifically (and religiously) advanced West is in the (usually urban) center, while Other cultures are in the wild, primitive, and underdeveloped (and undifferentiated) periphery. To reinforce this narrative, the achievements and contributions of colonized cultures are erased through disavowal, negation, and unacknowledged appropriation. Hybridity challenges this disavowal and appropriation by revealing the complexity of culture and cultural interactions and exposing the mutual dependency of the colonizer and the colonized and the way those contacts are internalized. It highlights the ways colonial cultures are the product of dynamic cultural interactions, of mutual borrowings. By revealing the ways that both colonizer and colonized are transformed, hybridity challenges center/periphery models of cultural interactions, as well as static and homogenous understandings which render ambiguity and difference invisible. PCSF, for example, often depicts the urban center not as a civilized space, but as one that is chaotic and dangerous. This urban center is not "pure" or orderly—it is a hybrid space, a mixture of races, species, technologies, dimensions. From the outside, they appear strange, chaotic, dangerous, and violent. This characterization is not without merit—they *are* dangerous—but they are also spaces of possibility of inclusive wholeness—a promise which is realized to varying degrees within the stories.[34]

The hybrid in PCSF is also an opportunity to resist erasure of colonial culture by revealing how indigenous culture exists within and contributes to colonial and postcolonial cultures, and by intentionally interacting with and transforming the culture of the colonizer into something new. The hybrid may look the same, but it is transformed by the process of displacement and relocation; their meanings are different, as are the meanings they produce.[35] The hybrid expresses the horror of colonialism, without succumbing to it. It is this quality that makes the hybrid ambiguous. Like Frankenstein's creation, it is monstrous because it is "uncannily both us and not us"; but, "wildly hopeful for the same reason." Jessica Langer describes this ambiguous hybrid as a "hopeful monster," reflecting the "simultaneous horror and potential" of the hybrid.[36] The hybrid is monstrous because of the trauma of colonization and the cultural wound created by "a *forced* collapse of boundaries by a strong and violent power, and the deliberate erasure of indigenous cultures,"[37] but it is simultaneously hopeful because it suggests the possibility of a future beyond annihilation. This hopefulness exists because the hybrid represents a strategy of subversive resistance, a way to avoid the destruction of indigenous culture, not through a return to a pristine past that represents the destruction of the colonizer and eradication of all traces of colonialism,[38] but by creating and describing new forms that synthesize disparate cultures and traditions. Here again, the hybrid describes a space for creating new cultural forms, an opportunity to create a new system of meaning from what has been imposed from the outside.

This chapter has attempted to demonstrate the usefulness of simultaneously considering three revolutions—that is, the relationship of religious, scientific, and global perspectives and not just the relationship of (Western) science and (Western) religion.[39] This helps reveal the ways religion, science, and colonialism are intertwined, particularly in what Jaffe describes as the structure of elective monotheism, which is not merely the impact of a single truth and a single interpretation, but the existence of an exclusive community responsible for creating a global community united by this single, universal truth. I suggested hybridity as an alternative to this static and monolithic understanding of truth and community. Hybrid thinking—and queer theories—can be helpful in a number of ways that go beyond the scope of this paper. Anzaldúa, for example, extends her understanding of borderlands thinking to a more ontological understanding of the relationship of spirit and matter—an aspect of her thought that is often ignored, but which would have strong resonances within the religion and science discourse, particularly among those who are process oriented. Bruno Latour and Anne Fausto-Sterling use the "hybrid" to describe the relationship between science and culture—hybrids are everywhere, "mixing ingredients of science, power and knowledge; economics, sex, and war"[40]—a position that pushes back against a narrow understanding of religion and science and suggests a kind of hybrid methodology—one that is truly *inter*disciplinary (not multidisciplinary). This also suggests an alternative to epistemology that more strongly resembles a critical approach developed by Sandra Harding, which relies on multiple standpoints instead of a single privileged position.

This understanding of hybridity that I have described above also provides critical insights into some of the limitations of the current approach to our understanding of religion and science, and suggests some of the ways we can be a more hybrid space. As long as we focus on singular truth claims—with the somewhat implicit assumption that Western, rational, and scientific perspectives are the best (or only) way to determine what "really" is, the only option will be to ignore, suppress, or exclude divergent perspectives, often in the interest of "finding agreement." Here, Anzaldúa's perspective is particularly helpful, and extending her insights into identity into a discussion of community, we might see difference is a kind of ontological given, which instead pushes us to greater inclusivity, consciously choosing to be more inclusive—seeking out alternative perspectives and challenges, and embracing the conflict, struggle, contradictions, and ambiguity that that entails. This process will be painful, but to those who worry about that, it is worth noting that the perspective of the global revolution reminds us that denying difference is also painful—and violent—the pain and violence is merely excluded from our frame of reference. Creating a space that is open to diversity does not make difference possible; difference exists, and it is this difference that makes it possible to be a diverse space. Inclusivity is merely an acknowledgment of the reality of difference. In this more inclusive space, instead of asking how we determine which claims are true, or how we might reconcile different truth claims, we would instead examine the mutuality of the cultural interactions that produce those truth claims—Who produces them and how? Who do they impact and how? Approached in this way, the relationship between multiple ways of knowing—including those that we call religious and scientific—becomes hybrid: a dynamic site of cultural production and creativity. This understanding makes it clear that the goal of being a more inclusive space is not a utopian vision, in which differences have been reconciled, because that kind of inclusivity requires suppression and erasure of difference, as well as the suppression and erasure of the suffering and injustice that is the price paid for utopia. The contestations of hybridity are not smooth or easy, and hybridity represents "a site of simultaneous despair and resistance," of destruction and creation. But, it is also "the key to a balance of sorts, . . . the *only possible way forward* in the current reality.[41]

WORKS CITED

Adas, Michael. *Machines as the Measure of Men: Science, Technology, and Ideologies of Western Dominance*. Ithaca: Cornell University Press, 1989.

Aldiss, Brian, with David Wingrove. *Trillion Year Spree: The History of Science Fiction*. London: Gollancz, 1986.

Anzaldúa, Gloria. *Borderlands/La Frontera: The New Mestiza*. San Francisco: Spinsters/Aunt Lute, 1987.

Bauman, Whitney A. *Religion and Ecology: Developing a Planetary Ethic*. New York: Columbia University Press, 2014.

Bell, Duncan. "Ideologies of Empire." In *The Oxford Handbook of Political Ideologies*, edited by Michael Freeden and Marc Stears. New York: Oxford University Press, 2013.

Bhabha, Homi. *The Location of Culture*. New York: Routledge, 1994.

Biggs, David Andrew. *Quagmire: Nation-Building and Nature in the Mekong Delta*. Seattle, WA: University of Washington Press, 2010.

Davis, Edward B. and Michael P. Winship. "Early Modern Protestantism." In *Science and Religion, A Historical Introduction*, edited by Gary B. Ferngren. Second edition. Baltimore, MD: Johns Hopkins University Press, 2017. 110–122.

Deason, G. B. "The Protestant Reformation and the Rise of Modern Science." *Scottish Journal of Theology* 38, no. 2 (May 1985): 221–240.

Dillenberger, John. *Protestant Thought and Natural Science: A Historical Study*. New York: Doubleday and Company, 1960.

Dillon, Grace L., ed. *Walking the Cloud: An Anthology of Indigenous Science Fiction*. Tucson, AZ : University of Arizona Press, 2012.

Fausto-Sterling, Anne. "Science Matters, Culture Matters." *Perspectives in Biology and Medicine*. 46, no.1 (Winter 2003):109–124.

Feghali, Zalfa. "Re-articulating the New Mestiza." *Journal of International Women's Studies* 12 no. 2 (2012): 61–74.

Fotopoulou, Aristea. "Intersectionality, Queer Studies and Hybridity: Methodological Frameworks for Social Research." *Journal of International Women's Studies* 13, no. 2/3 (2012): 19–32.

Harding, Sandra. *Is Science Multi-Cultural? Postcolonialisms, Feminisms, and Epistemologies*. Indiana University Press: Bloomington, 1998.

Hopkinson, Nalo and Upinder Mehan, eds. *So Long Been Dreaming: Postcolonial Science Fiction and Fantasy*. Vancouver, BC: Arsenal Pulp Press, 2004.

Jackson, Peter A. "Global Queering and Global Queer Theory: Thai [Trans]genders and [Homo]sexualities in World History." *Autrepart* 49, no.1 (2009): 15–30.

Jaffe, Martin S. "One God, One Revelation, One People: On the Symbolic Structure of Elective Monotheism," *Journal of the American Academy of Religion* 69, no 4 (December 2001): 753–775.

Langer, Jessica. *Post Colonialism and Science Fiction*. London: Palgrave Macmillan, 2011.

Latour, Bruno. *We Have Never Been Modern*. Cambridge, MA: Harvard University Press, 1993.

McClellan, James E. III. *Colonialism and Science: Saint Domingue and the Old Regime*. Chicago, University of Chicago Press, 1992.

Merchant, Carolyn. *The Death of Nature: Women, Ecology and the Scientific Revolution*. New York: HarperCollins, 1980.

Muñoz, José Esteban. *Disidentifications: Queers of Color and the Performance of Politics*. Minneapolis: University of Minnesota Press, 1999.

Ortega, Marianna. *In Between: Latina Feminists Phenomenology, Multiplicity and the Self*. Albany: State of New York Press, 2016.

Rieder, John. *Colonialism and the Emergence of Science Fiction*. Middletown, CT: Wesleyan University, 2008.

Said, Edward. *Culture and Imperialism*. New York: Vintage Books, 1993.

Schiebinger, Londa. *Plants and Empire: Colonial Bioprospecting in the Atlantic World*. Cambridge, MA: Harvard University Press, 2004.

Smith, Andrea. "Queer Theory and Native Studies: The Heteronormativity of Settler Colonialism." *GLQ: A Journal of Lesbian and Gay Studies* 16, no. 1–2 (2010): 42–68.

Spurlin, Wiliam J. "Sexual/Cultural Hybridity in the 'New' South Africa Emergent Sites of Transnational Queer Politics." In *Gendering Border Studies*, edited by Jane Aaron, Henrice Altink, and Chris Weedon. Cardiff: University of Wales Press, 2010.

Warner, Michael. "Introduction: Fear of a Queer Planet." *Social Text*. 29 (1991): 3–17.

NOTES

1. Martin S. Jaffe, "One God, One Revelation, One People: On the Symbolic Structure of Elective Monotheism," *Journal of the American Academy of Religion*, December 2001, 69, no 4:753–775.

2. Edward Said, *Culture and Imperialism* (New York: Vintage Books, 1993), 9.

3. Duncan Bell, "Ideologies of Empire" in *The Oxford Handbook of Political Ideologies*, ed. by Michael Freeden and Marc Stears (New York: Oxford University Press, 2013), 538–43.

4. John Rieder uses this term, building on Slavoj Žižak's concept of ideological phantasies—"beliefs that we consciously disavow, recognizing them as untrue, but nonetheless support in practice"—to elaborate on a number of ideological fantasies associated with colonialism (John Rieder, *Colonialism and the Emergence of Science Fiction* [Middletown, CT: Wesleyan University, 2008], 29–33).

5. James E. McClellan, III, *Colonialism and Science: Saint Domingue and the Old Regime* (Chicago: University of Chicago Press, 1992), 7. See also, Michael Adas, *Machines as the Measure of Men: Science, Technology, and Ideologies of Western Dominance* (Ithaca: Cornell University Press, 1989), Sandra Harding, *Is Science Multi-Cultural? Postcolonialisms, Feminisms, and Epistemologies* (Indiana University Press: Bloomington, 1998), Londa Schiebinger, *Plants and Empire: Colonial Bioprospecting in the Atlantic World* (Cambridge, Harvard University Press, 2004).

6. Adas, 2.

7. E.g., San Dominigue (McClellan, 8–9) and the Mekong Delta (David Andrew Biggs, *Quagmire: Nation-Building and Nature in the Mekong Delta* [Seattle, WA: University of Washington Press; 2010], 48–50).

8. McClellan, 123.

9. McClellan, 136; Schiebinger describes in detail how knowledge of abortifacients, such as the peacock flower, were controlled and access restricted not only from native populations and slaves (who would abort to prevent a child from being born into slavery), but also from European populations. 105–193.

10. Adas, 4.

11. Adas, 3–4.

12. Adas, 22.

13. Adas, 7.

14. Harding, 204.

15. Adas, 205. For a discussion of the role of institutional religion in the development of cochineal and quinine, see Schiebinger 38–44 and 214–127.

16. Adas, 205–07.
17. Adas, 6.
18. Jaffe, 771.
19. Jaffe, 754.
20. Harding, 3. See also, Whitney A. Bauman, *Religion and Ecology: Developing a Planetary Ethic* (New York: Columbia University Press, 2014): "This is same sort of foundationalism gets taken up into understandings of scientific progress (replace revealed creation with natural laws and revelation with reason) and in ideas of economic development (replace fallen creation with primitive, and salvation with civilization)." (20).
21. See Introduction, p. 22–24. I would also add feminist perspectives, critical race theory, and other critical discourses. For a broader discussion of the connections and blind spots between these trajectories, see Aristea Fotopoulou, "Intersectionality, Queer Studies and Hybridity: Methodological Frameworks for Social Research," *Journal of International Women's Studies* 13, no. 2 (2012): 19–32. Andrea Smith, "Queer Theory and Native Studies: The Heteronormativity of Settler Colonialism," *GLQ: A Journal of Lesbian and Gay Studies* 16, no.1–2 (2010): 42–68. Michael Warner, "Introduction: Fear of a Queer Planet," *Social Text* 29(1991): 3–17.
22. Gloria Anzaldúa, *Borderlands/La Frontera: The New Mestiza* (San Francisco: Spinsters/Aunt Lute, 1987), 102.
23. Anzaldúa, 106, 102–03.
24. Anzaldúa, 100.
25. Anzaldúa, 104–105.
26. Anzaldúa, 99.
27. Anzaldúa, 102.
28. See, e.g., José Esteban Muñoz. *Disidentifications: Queers of Color and the Performance of Politics* (Minneapolis: University of Minnesota Press, 1999); Smith, 55–58.
29. Wiliam J. Spurlin, "Sexual/Cultural Hybridity in the 'New' South Africa: Emergent Sites of Transnational Queer Politics," in *Gendering Border Studies*, ed. Jane Aaron, Henrice Altink, and Chris Weedon (Cardiff: University of Wales Press, 2010), 215.
30. Jessica Langer, *Post Colonialism and Science Fiction* (London: Palgrave Macmillan, 2011), 107.
31. I use the term speculative fiction, rather than science fiction, because narrow interpretations of "science" are often used to exclude women and people of color from the canon of "serious" science fiction. "Speculative fiction" refers to fiction which "occurs in a place that never existed, on a world we know nothing of, or on an earth that might have been or might be" (Robert Heinlein, "On the Writing of Speculative Fiction" in *Of Worlds Beyond; The Science of Science Fiction Writing*, Lloyd Arthur Eshbach, editor [Chicago: Advent Publishers, 1964], 2). It includes science fiction, but also magical realism, alternative history, and apocalyptic, utopian, and dystopian literature.
32. *Walking the Cloud: An Anthology of Indigenous Science Fiction*. Grace L. Dillon, editor. Tucson, AZ: University of Arizona Press, 2012, 2 and 7.

33. Langer, 10.

34. See, *Brown Girl in the Ring* and *Perdido Street Station* as well as this citation to the work that looks at them.

35. Homi Bhabha, *The Location of Culture* (New York: Routledge, 1994), 172.

36. Langer, 107.

37. Langer, 110.

38. Langer, 108.

39. And it should be noted that there will also be additional perspectives—additional revolutions, if you will—to include, each of which will broaden and deepen our understanding.

40. Anne Fausto-Sterling, "Science Matters, Culture Matters," *Perspectives in Biology and Medicine* 46, no.1 (Winter 2003):112.

41. Anzaldúa, 109.

Chapter Four

Queering Authority in Science and Religion

Whitney Bauman

The premise of this chapter is that transformative knowledge comes about through unknowing. In her recent book, Catherine Keller traces the "Cloud of the Impossible" in the Medieval Theologian Nicholas of Cusa and makes connections between Cusa, Bruno, Quantum Physics, and even deconstructive philosophies.[1] In order to create spaces in which new ideas might emerge, these thinkers and concepts help us to take stock of the more problematic aspects of claims of certainty. Similarly, Karen Barad blurs the boundary between the proposed great divide between ontology and epistemology through her engagement with the differences between Bohr and Heisenberg.[2] Finally, Jack Halberstam traces a type of unknowing in *The Queer Art of Failure*. Halberstam notes that if the current model of "production" is leading us toward eco-social destruction and planetary catastrophe, then perhaps the best thing we can do is to cultivate failure.[3] In these three models we find a method for a different way of understanding epistemology that might be called experimental worlding. The first part of this chapter, then, will examine the common ground between these three thinkers: first, that received and established knowledge helps us to order the world in a preconceived way in which newness becomes interpreted as more of the same; and second, that there is something about unknowing, uncertainty, and even failure that helps us to cultivate a form listening in which the voices of previously unheard planetary realities come into speech and can then contribute toward new ways of being/becoming in the world. According to this perspective, when we look at shifts in knowledge that transform the worlds in which we live, these shifts come about as a result of paying attention to that which is made abject by received traditions. The second half of this brief chapter will look at three examples of this method for unknowing: the Pope and Climate Change, Marjory Stoneman Douglas, and Rachel Carson. I'm not suggesting that these three people are queer theorists or activists, but they do participate in queering the boundaries between science and religion, epistemology and ontology, humans and the rest of the natural world. More importantly, their authority within their respective roles and fields comes in large part through the unknowing and unlearning of received knowledge that each of them had to go through, in order for a new understanding to emerge which took account of religious, scientific, and other aspects of their contexts. In other words, their authority came from their ability to listen deeply to the multiple voices, human and other, in their given contexts.

THE PERILS OF AUTHORITY IN SCIENCE AND RELIGION: TOWARD EXPERIMENTAL WORLDING (UNDOING AND RE-MEMBERING)

> This is a dream not of a common language, but of a powerful infidel heteroglossia. It is an imagination of a feminist speaking in tongues to strike fear into the circuits of the supersavers of the new right. It means both building and destroying machines, identities, categories, relationships, space stories. Though both are bound in the spiral dance, I would rather be a cyborg than a goddess.[4]

In Haraway's classic essay on the "Cyborg Manifesto," she takes aim at authority claims wherever she finds them. In this specific case we have on the one hand the male-dominated metaphors of mechanical science and on the other the female-dominated metaphors of organism encapsulated in the idea of Mother Earth. She sees in these two metaphors similarity in the way that authority is claimed. On the one hand it is through a cold, detached (impossible) objectivity from which we can manipulate the world. On the other hand it is through a deep embodied intuitive connection, from which we can again make ordering claims about the world. In both cases, in other words, gender and race relations among humans, relations between humans and animals, and relations between organic and machine are concretely misplaced in ways that do not actually make sense of the hybrid, relational, and changing worlds in which we live. In doing so, all sorts of normative lines can be drawn that do not serve any sort of dialogue but rather serve as reinforcements for maintaining a specific view of the world. Both options, in other words, claim some sort of secret (impossible) knowledge that is authoritative. Both are couched in impossible certainty.

Here I would like to argue for a method for dialogue and knowledge production which opts not for certainty, but rather for uncertainty. The outcome of such a method is purposefully not already decided because such decisions foreclose on unforeseen possibilities. Rather, values emerge from the process itself, each time such a process is applied. First, let me describe by way of Karen Barad, Catherine Keller, and Jack Halberstam why I think the blurring of epistemology-ontology, the art of unknowing, and the art of failure (respectively) are all three important and then I will move on to discuss some of the features of this experimental worlding process.

Karen Barad in her book *Meeting the Universe Halfway* extends the idea of performativity to the quantum level. In other words, in a somewhat Whiteheadian manner, she suggests that there is some level of "agency" at all levels of reality. Such agency, what she calls agential realism, means that there is no one way in which the world should, ought, or was made to be.[5] Thus, the question is not one of how our epistemology "fits" the world out there (ontology); rather the questions we pose to the world actually help shape the worlds in which we become. Of course we are talking about this from a human-perspective, but this type of "sensing" of the world is happening at all levels of life.

What does this mean? For her it means following Bohr's idea of ontological indeterminacy rather than merely Heisenberg's theory of epistemological uncertainty, for starters. But more importantly for us it means that knowledge is a co-production of the world, by various entities. As Barad notes in one article, "Phenomena are entanglements of spacetimemattering, not in the colloquial sense of a connection or intertwining of individual entities, but rather in the technical sense of 'quantum entaglements' which are the (ontological) inseparability of agentially intra-acting 'components.'"[6] The more entities that are intentionally involved in our process of knowledge production, ergo, the more possibilities for becoming we will have. Again, she notes, "responsible laboratory practices must take account of the agential performances of the organism in making the specific nature of causal relations evident. *Responsibility entails providing opportunities for the organism to respond.* Dinos do not respond to deterministic models of causality."[7] Such a method is hard to follow because we are lured toward individual certainty in knowledge claims rather than an understanding of what we don't know, or some sort of multi-perspectival approach to knowledge production.

Catherine Keller, drawing from Barad among many others, traces what might be called the art of unknowing from Nicholas of Cusa, through Bruno, other apophatic and negative theologians, and deconstructive and post-structural philosophies.[8] What this journey through the fecund darkness of our projected certainties uncovers is that much newness emerges not from the spaces of projected certainty but from the covered over or ignored spaces of uncertainty. In other words, what we don't know is often what leads toward creative transformations in the world (more on this idea will be shared below).

If what we don't know is what often leads us to creative transformation then what we do know often leads us astray. Wait! We are topsy-turvy! And this is exactly the point at which failure becomes an art. When we are, as Jack Halberstam notes, participating eagerly in a habitual knowledge that is destroying many planetary bodies (human and other), then we need to participate in the art of failure.[9] Failing in the face of what progress means in a world where progress leads to gross economic inequity, climate weirding, and countless other socio-ecological disasters becomes a transformative practice. From these three thinkers, we can begin to articulate an experimental worlding method, which includes scientific, religious, and other information in an effort to open us onto a multitude of planetary partners. There are at least four parts to this method.

First, we begin to understand our own selves as agents among agents: as what Deleuze and Guattari, and others following them, call assemblages.[10] We understand that we are conglomerations of flows of animals, materials, minerals, languages, histories, and cultures. At each level, there is some action or determination of what we call our "self," and at each level we also add layers to that understanding through our languages, histories, and cultures. We are at the same time multiple and interrelated with the rest of the evolving planet, and much less powerful than we might imagine from within a monotheo-influenced culture—hints of polydoxy and multiplicity here intended.[11] If theology becomes anthropology (or vice versa) then we need to think with the multiplicity of the trinity (perhaps) or polytheism more so than with the omni-God of logocentrism and single origins.

Second, if we are more in tune with multiplicity, we can begin to understand ourselves as never fully there, but always of multiple parts and contributing to a multi-logue of voices. It is, from this understanding, the self or unity that is an abstraction, while the multiplicity is the lived reality.[12] Whatever we might understand from our own perspective is just *a* perspective along the way: even if it is the common/normative perspective of human beings, mammals, or earthlings. There is common ground in our speech amidst this multiplicity—we do all, for example speak from a common planet—but we can never capture all that is in these multiply evolving bodies with language. There are always remainders or abjections.[13] Whatever knowledge is, it must take account of these multiple remainders and abjections. This is, as Haraway and others have noted, a way out of the false binary of universalism and relativity: viz. contextuality.[14] Contextuality, paying attention to more and more perspectives in any given situation, phenomena, and event, becomes the new grounds for what we might think of as "objectivity," or what Sandra Harding has called "strong objectivity."[15] None has a full perspective because this would be the disembodied perspective from nowhere, but each adds to the overall understanding of a given situation.

Through attention to the abjections and remainders of our knowledge claims, we can begin to collect the many unheard voices of our collectivity.[16] This is perhaps best done (as we will see below) through deep listening and sharing of stories. As Lisa Stenmark points out, rightly I think, the importance of narratives in our systems of knowledge—scientific and religious together—cannot be underestimated.[17] If we stick to one story, we lose a lot of meaning and this is true whether we are talking about eyewitness accounts of a crime, scientific observations of a phenomenon, or what a specific event or person means to another. Knowledge is more about the proliferation of meanings than a narrowing of meaning.

This evolving multiplicity is important for the fourth and final point, which involves a mapping of all of these multiple ways of knowing so that we might better understand the many different possibilities for collectively becoming into the planetary future. From where we stand, the present and personal perspective seems obvious; but from the future, such perspectives seem quaint and located. Just think of how ideas from the past seem odd from current scientific and religious perspectives, and such will be the case with our own ideas in the future. Unless there is some magical transition that happens during the continual movement of present moments becoming past, then I would wager that our own present perspectives do not hold all there is to know any more than any other present moment has. We should not mistake our feelings, as Whitehead might quip, for concreteness.[18] Further, if our world truly is multiperspectival and interrelated in some way, then hearing the many voices (human and non) into speech about the ways we might inhabit the future becomes a question of ethics and aesthetics: what types of worlds do we want to co-create and what types of justice and injustice (toward earth bodies) come along with these worlds.[19] Again, as Barad notes, "Differentiating is a matter of entanglement. Entanglements are not intertwining's of separate entities but rather irreducible relations of responsibility. There is no fixed dividing line between 'self' and 'other,' 'past' and 'present' and future,' 'here' and 'now,' 'cause' and 'effect.'"[20] With such an understanding of reality, and our knowledge of it, we might then be able to begin mapping the possibilities of our becoming onto earth-bodies, and the different effects such possibilities have on the plethora of earth bodies. And, with these mappings we can decide which ways we think are better and worse for the planetary future. This type of mapping suggests an authority based upon unknowing, and trust in the multiply embodied perspectives of earth others. It weaves together a common ground for knowing rather than projecting knowledge onto uncertainty.

These four points contribute to what I would call an experimental worlding process. If we were to take our worlding processes and treat them as, well, experimental, then we might be able to better map the effects of these processes and their effects on earth bodies in a much more accurate way that helps us to respond to the changing conditions and needs of the planetary community. Again, this is not about willy-nilly *deciding* how we are going to create words as if *ex nihilo*, but rather it is about focusing our meaning-making practices on their ethical and aesthetic effects rather than on their metaphysical truths. The shift in focus outlined above, then, might help us to shift categorical boundaries—male/female, animal/human, science/religion—in ways that might better address the needs of the always porous boundaries within the becoming planetary community. What I want to do with the remainder of this chapter, then, is focus on some examples of such a method of unknowing in some seemingly unlikely peoples and places.

THE GIFTS OF UNKNOWING: THE GREEN POPE, THE GRANDE DAME, AND THE SEA AROUND US

Though the Pope, Marjory Stoneman Douglas, and Rachel Carson are not exactly what comes to mind when one thinks of the word "queer," there is much to be commended in their approaches to knowing and foundations of knowing that I would argue suggests they each have a queer epistemology. All three of them are, first and foremost, cobbling together information from religion and science (and other sources) into a hybrid understanding of very contextual issues (planetary climate change, South Florida and the Everglades, and industrial agriculture, respectively). In doing so, they blur the boundaries of received knowledge that falls under the authority of "religion" and under the authority of "science." This can be seen in the way the Pope dabbles in scientific information, Douglas (a journalist) in environmental sciences, and Carson (a scientist) in literature and ethics. In addition to being deeply engaged in contextual multiperspectivalism and the art of unknowing, all three figures have what might also be called "queer" understandings of personal relationships (at least non-heteronormative): celibate, single, and lesbian. These are not just trite points: if heteronormativity helps keep us locked into certain ways of relating to other humans and the rest of the natural world, then living in a non-heteronormative way might lead to different ways of understanding these relationships. Thought I am not arguing that these three figures are "queer" in terms of being queer theorists or even self-identifying as such, I am arguing that they each offer elements of what queering authority in knowledge claims might look like.

The Green Pope

There is much to be commended in the Pope's recent Encyclical on "the environment." He clearly did his doctrinal, historical, and philosophical homework on issues of human-earth relations. There is much one would expect to find in the document: such as the reference to St. Francis in the title, couching of creation-care in terms of "the common good" and the prominence of catholic social teaching. There are also some surprises—such as his knowledge of the history of the environmental movement and his use of Integral Ecology, which understands nature and culture as already and always together. At times it reads like a traditional Papal document while at others it reads more like something that Bruno Latour or scholars of the New Materialism might have written.[21] Here, I focus on two points that I find most poignant in the Encyclical: the critique of modern technological society and the call for a more robust dialogue between religion and science. Both of these points participate in what Jack Halberstam calls "The Queer Art of Failure."

In *The Queer Art of Failure*, Halberstam writes:

> Under certain circumstances failing, losing, forgetting, unmaking, undoing, unbecoming, not knowing may in fact offer more creative, more cooperative, more surprising ways of being in the world.[22]

The creativity of abject identities, of those who have failed to live up to the norms of hetero-normative or homo-normative, anthropocentric capitalism, is indeed the source of creativity for seeking a different planetary future. In other words, failing is precisely what we need in this day and age if we are to find our way forward through the problems brought about by globalization and climate weirding. Pope Francis hints at this in his critique of modernity found in the Encyclical. He writes (and here I quote at length):

> §107. It can be said that many problems of today's world stem from the tendency, at times unconscious, to make the method and aims of science and technology an epistemological paradigm which shapes the lives of individuals and the workings of society. The effects of imposing this model on reality as a whole, human and social, are seen in the deterioration of the environment, but this is just one sign of a reductionism which affects every aspect of human and social life. We have to accept that technological products are not neutral, for they create a framework which ends up conditioning lifestyles and shaping social possibilities along the lines dictated by the interests of certain powerful groups. Decisions which may seem purely instrumental are in reality decisions about the kind of society we want to build.

What the Pope is critiquing then is the current "productionist paradigm" which is being exported through western dominated globalization as itself "natural."[23] The current measure for individual success and for the "development" of other countries is literally leading to a world in which many of our lives are so sped up and removed from their eco-social consequences that we are outstripping the carrying capacity of the planet.[24] Some prefer to name this the anthropocene but it might be more aptly called the fossil-fueled capitalocene,[25] or along with Haraway, the cthulucene, which takes into account our entanglement with other earth-creatures and processes.[26] On the one hand, not all humans are equally guilty for contemporary eco-social crises, and on the other hand human begins do not hold all agency for solving the world's problems. The world, after all, is not dead matter—as the Pope readily argues in his Encyclical.

In an important sense the Pope is arguing that in light of this context we all need to practice "failure": or that which disrupts the "business as usual" notion of progress as solely economic and technological. I'm not suggesting this Pope is queer-friendly (or even feminist-friendly), but I am suggesting that the deeply Catholic understandings of the "common good" and "social teachings" are, in the face of the productionist paradigm, queer. To further add to this failure, he then argues for a type of unknowing, or epistemic humility in the relationship between religion and science.

Again, he writes:

> §199. It cannot be maintained that empirical science provides a complete explanation of life, the interplay of all creatures and the whole of reality. This would be to breach the limits imposed by its own methodology. If we reason only within the confines of the latter, little room would be left for aesthetic sensibility, poetry, or even reason's ability to grasp the ultimate meaning and purpose of things.

The suggestion that values and religious ideas ought to be taken as serious as science may make some climate scientists and fundamental atheists of the Dawkins stripe cringe. But this statement stands firm when one considers that we have had the science and technology to deal with ecological problems for years, yet these problems just keep getting worse. The position the Pope argues for in the Encyclical exemplified in the above passage is one that depends heavily on the apophatic and negative traditions within theology in at least two ways.

First, he is doing much in this Encyclical to undo the seemingly solid and separate terrains of scientific and religious ways of knowing. Many have argued that a large portion of the blame for our eco-crises lies at the feet of religions which promote forms of human exceptionalism, if not domination.[27] While this is partially true, it does not mean that science and technology do not suffer from the same pitfalls. They were, after all, in large part reared in the monotheistic cultures that they want to claim a complete separation from. The notion of a single good creation easily morphs into the notion of a nature based upon universal natural laws; the notion of humans as having dominion over or being "special" easily morphs into the idea that all non-human life can be manipulated via science and technology; and the notion that we will be saved from the cycles of predator/prey, life/death, morphs easily into the idea that science and technologies will save us and help us outsmart our creaturely, embodied fate. In the end, science and religion may have different methods for approaching the world and one may produce more persuasive results (here and now), but they are both human adventures and they are historically always already tied to one another.

Second, and related, he is asking for epistemic humility in current dialogues between multiple parties as we search for possibilities for future becoming. There is no extant form of human thinking that was created during a time in which we are dealing with global climate change and the "acceleration" caused by globalization.[28] To claim that we might be able to know the answers means that we are projecting the past onto the future. Rather than participate in this form of colonizing, we might enter into a space of unknowning, which allows for multiple voices to be heard, and thus unthinkable possibilities for future becoming to emerge. To quote from Catherine Keller's *Cloud of the Impossible*, this is a move from "the unspoken and the unknown tilt from the hierarchical verticality of the Neoplatonic being beyond being, toward the horizon of our collective convulsions"[29] (albeit from the head figure of the hierarchical Catholic church).

The Grande Dame[30]

Yet another example of the fecundity of unknowing and queering boundaries, can be found in the Grande Dame of the Everglades, Marjory Stoneman Douglas. Marjory Stoneman Douglas is almost synonymous with the history of the formation of Miami and the Everglades National Park. With a life spanning 108 years (1890–1998), she was witness to the effects of the Industrial Revolution, two world wars and the rise of nuclear threats, the environmental movement in the United States, and the fall of communism and rise of the post-cold war world in which we now live. Though regarded by many as just as important as Rachel Carson, John Muir, and Aldo Leopold in terms of her environmental advocacy, she is far less known.

Her brand of regionalism as environmentalism provides grounds for a narrative that bridges environmental justice and traditional (white and middle to upper class) American understandings of environmentalism. In other words, she provides what we might call an early example of a planetary ethic that bridges the social and the ecological, peoples and nature, and the urban and the wild. Such an ethic is helpful in an era of globalization and climate change, when these categories are in flux.

Though a self-proclaimed agnostic, Marjory Stoneman Douglas came from a long line of Quakers and abolitionists. It was perhaps this social justice oriented way of understanding the world that led her quickly to see the problems of "environmental justice" in the early development of Miami and its relationship to the Miccosukee Indians and the Everglades. It is at least plausible to suggest that her background Quakerism aided her understanding of the connection between urban and wild, and between the human and nonhuman world. In other words, the leanings of Quakerism toward non-hierarchical and democratic thinking promotes an understanding of life that is amenable to hearing the many voices: and for Douglas, this also meant the voices of the nonhuman world.[31] She articulates what might be called a multiperspectival and pluralistic environmental ethic. As a result of this orientation, she found herself on many different sides of environmental arguments: from being on the side of the Corps of Engineers in helping to better engineer the "river of grass"; to being against the use of plumage in 1920s/30s fashion, which decimated the avian population of the Everglades; to fighting against Big Sugar once she realized the destruction that it wreaked on the flora and fauna of the Everglades. Douglas methodically weighed human and nonhuman interests and came out on the side that she thought would promote the most good for the entire community. Douglas had an ability to, in Nelle Morton's terms, hear voices into speech.

> Hearing to speech is political. Hearing to speech is never one-sided. Once a person is heard to speech she becomes a hearing person.[32]

To read the biography and the writings of Marjory Stoneman Douglas is to understand her as an exemplary model of the method of "hearing voices into speech." This in no small way is related to her process of unlearning and unknowing. As a woman in science dominated by a man's world, and as someone who was challenging economics as usual with her understanding of the relationship between humans and nature, she had to face a great deal of criticism from those who policed the received scientific and political canon.

For starters, she challenged the idea of European erasure of the indigenous peoples of the Everglades.

> To her, the tragic drama of powerful people dominating both less powerful people and compliant nature ran like an unbroken thread throughout the history of Western civilization, with repeat performances played out in the Everglades. Her insights here prefigured the idea of environmental justice, which came to life comparatively late in the environmental century.[33]

This was a powerful narrative, in other words, that fell in line with the received knowledge that western civilization created useful cultivation through expansion and takeover of lands (which were otherwise "empty") until that point. In order to unlearn this knowledge, Douglas had to be willing to listen to the many voices making up the unique histories and landscapes of South Florida. She also tied together here issues of social injustice and ecological degradation. She challenged the language of the Everglades as a dark and scary swamp (which harbored runaway "Seminoles") and argued both that they had made a living and culture in this place, and that it was indeed a living "River of Grass."[34]

Not only did this unknowing lead to changes in the thinking about the Everglades and human relationships to the Everglades as something that would become a national treasure rather than something that should be "drained" and made useful, but her unknowing spilled over into other received and related institutions as well. For instance, she was a critic of the enshrining of the American ideal of the family, which of course is also critiqued as one of the pillars holding up a capitalist society.

> She questioned the importance of a sacred but frail American institution: the family. Many of society's ills grew out of dysfunctional relationships at home, she honestly claimed. Properly run orphanages provided a more salutary environment than some families she had known.[35]

Finally, she also had to come to terms with unlearning her trust in the government, yet another form of unlearning that required hearing voices that were left out of official stories about human-human and human-earth relations.

Far from not knowing anything at all, it is precisely in "unknowing" that new relationships to knowledge and bodies are forged and formed. In other words, unknowing our knowledge-systems, our truth-regimes, provides more possibilities for becoming in different ways that are up until that point, unforeseen. Yet another pioneer in this sense is the final person I want to discuss, viz. Rachel Carson.

Rachel Carson: Wonder, Humility, and Sea Ethics

> Carson's sea ethic is perhaps most apparent not in what she wrote in *The Sea Around Us* but in what she chose to omit.[36]

Just as Douglas's relationship with the Everglades served to trouble any type of reductive understanding of environmental management or environmental ethics, and instead blurred the boundaries between urban and wild and humans and the rest of the natural world, Carson's relationship with the sea also provided fodder for an ethic of ambiguity. "Carson portrays the ocean as vast, indomitable, and omnipotent. Humans, by contrast are depicted as small and somewhat insignificant, as unable to control and master the ocean's power and to harness it for our own ends."[37] Though she does not develop a full blown "sea ethic," it can be said that her relationship to and reflections on the sea have several components of what a queer understanding of "religion and science" might look like. Two components of such an ethic that I highlight here are her critique of human mastery and the importance of fostering a sense of wonder. Neither of these suggests that she was not adamant about gathering scientific data: she was, after all, a scientist. However her scientific inquiry was couched in a wider sense of mystery and this is key to her appreciation for the rest of the natural world. Carson, though agnostic and raised in a Presbyterian family, might today be what is called a religious naturalist. In other words, the mystery of the natural world provided for her a wider sense of value and purpose in life. Regardless of whether she would adopt this label or not, she was influenced by Albert Schweitzer's "reverence for life," and such reverence does show up in her delicate balance between objective scientific observations and respect for the mysteries of the rest of the natural world.[38]

Though we may look back on her life now and see her in a single narrative line as one of the grandmothers of the modern environmental movement, the idea that the chemicals we used in agriculture and other parts of our daily lives could drastically change our environment and have repercussions for our own health was not widely of concern. This was the era of the Green Revolution, after all, and technology and science were supposed to save us from the provisions lacking in nature. Furthermore, the idea that we might cause enough harm to our fisheries and other ecosystems to cause their collapse was not one that was widely held or even thought about. Though there were dissenting voices, many scientists and intellectuals questioned whether there might be "limits to growth" at the time Carson was writing. It wasn't until a bit later that the infamous Club of Rome report was published.[39] Thus, her work entailed bucking the dominant ethos of the scientific community as well as the wider society. Much like Barbara McClintock and her genetic discoveries based upon a more wholistic (and less reductive) view of corn, so Rachel Carson's breakthrough came in part through her "sea ethic" of what might be called deep listening and humility.

This was an ethic that entailed a great deal of unknowing of received knowledge policed by the institution of mechanistic science. Hoewever, just because she was critical of the mechanistic view did not mean that she completely embraced a feminist organic view. Rather, she blurred the boundaries between these two models.

> As Carson incorporated the human body into "ecology," she also began to resolve a longstanding tension between two distinct narrative voices with which she had experimented in her major works on the sea: the "heavily masculine" voice of the scientist who coolly organizes, interprets, and summarizes reams of information, and the more feminine voice of the "appreciative nature writer," the close observer and participant, expressing wonder and enchantment with nature's mysteries.[40]

In a way we might say that Carson participated in a queering of the perspective of the relationship between humans and nature, and in doing so challenged the gendered stereotypes founded within that distinction. I'm not suggesting here that Carson was any sort of queer theorist, but rather that she unknowingly destabilized the ways in which gendered views of nature were used to anchor gender norms simply by destabilizing the human-earth relationship. In doing so, she challenged both the way in which established science "knew" nature and what "proper" male and female outlooks on the world might be. And, of course, she paid heavily for this as established science (and in particular those tied to agro-industry) fought her arguments about chemicals through attacking her intellect and personal character. The method by which she comes to this blurring is, perhaps, best articulated in her "sea ethic."

Carson sought to understand the life of the sea from the perspective of organisms that inhabited the sea: the coastal areas the shallow waters, and even the "abyss." Exploring the life of the deep sea was very much a new area at the time when she was writing about the sea. Accordingly, she attempted in her writing and research to take on multiple perspectives of sea life, and resisted the totalizing tendency in scientific knowledge to explain everything as fact. In other words, as the Pope took multiple perspectives into account with his *Green Encyclical*, and Marjory Stoneman Douglas had sought to "hear the many voices" into speech, so too with Rachel Carson. One might say that she allowed the contours of the ocean to change the way that she thought about scientific observations. She developed, in other words, what Lisa Stenmark has called "an ecology of knowledge."[41] A couple of examples might suffice. First, Carson writes,

> It is not the distance from the shore, but the depth, that marks the transition to the true sea; for wherever the gently sloping sea bottom feels the weight of a hundred fathoms of water above it, suddenly it begins to fall away in escarpments and steep palisades, descending abruptly from twilight into darkness.[42]

This is a typical statement from Carson: part scientific observation, part wonder, and part recognition of the limits of human knowing. Such limits provide the habitat for new insights "from the deep" to emerge.[43] In yet another blurring of boundaries, she articulates a type of deep understanding of the interconnectedness of all life and of the multiplicity of subjectivity. While observing the life-cycle of the mackerel, she writes,

> Of the millions of mackerel eggs drifting along the one . . . [viable egg] . . . thousands went no farther than the first stages of the journey into life until they were seized and eaten by the comb jellies, to be speedily converted into the watery tissue of their foes and in this reincarnation to roam the sea, preying on their own kind.

This observation is indicative, again, of the way in which she sees life as interconnected and in the way in which she understands the mystery/wonder of life. Again, she was influenced by Schweitzer's "reverence for life" which was in turn influenced by vedic cultures and their more relational understandings of the world. It is not too much of a stretch, then, to understand Carson in a long line of scientists who were re-thinking nature from a non-substantial based metaphysic. Such a metaphysic would fly in the face of a certain type of monotheistic understanding of nature, humans, and the world in which knowledge and reality were in a fixed and knowable relationship. Again, this is why I think her work was so threatening to so many. Yes, there were economic implications but more deeply her work further unraveled the frayed edges of a metaphysical system that ordered the western relationship between humans, knowledge, and the natural world.

Her sense of wonder sparked by learning about and listening to the rest of the natural world, perhaps kept her going in the face of her opponents. Just as Darwin and Einstein were afraid of the consequences of their own observations and didn't want to publish some of their more controversial results that bucked the common scientific ethos of their times, so with Carson. However, it was precisely this uneasiness with her own conclusions and her ability to be humble where many scientists projected certainty, that led to her own discoveries about DDT and its accumulative effects on our environments, not to mention our bodies. Yet again, the embrace of unknowing led to some very rapid changes to the ways in which we think about the world and some very socially and ecologically "progressive" benefits for the planetary community. Carson's witness and more importantly her humility have literally transformed the world in ways that allow for many more planetary voices and connections to come to speech.

In Carson's sea ethic, Douglas's understanding of the Everglades and Miami functioning as a single organism, and in the Pope's religiously and scientifically informed Encyclical on the environment, I think we find good models for queering religious and scientific authority in a post-secular age. We are, in our contemporary era and perhaps humanity is in every era, facing a future of uncertainty. It is not as easy as applying past knowledge categories onto the future becoming of life and thereby forcing a repeated sameness onto future generations. This is the quick and easy form of reasoning and working through problems: we make solutions based upon what we know. It is perhaps better to undo our categories of knowledge production such as what is classified as "scientific" and what is classified as "religious," participate in a bit of unknowing of received traditions, and provide future generations with the resources necessary to deal with the problems that we create by thinking creatively anew about our understanding of the planet and our places in it.

WORKS CITED

Barad, Karen. "Nature's Queer Performativity." *Kvinder, Køn & Forskning* 12 (2012): 25–53.
———. *Meeting the Universe Halfway: Quantum Physics and the Entanglement of Matter and Meaning*. Durham, NC: Duke University Press, 2007.
Bauman, Whitney. "Climate Weirding and Queering Nature: Getting Beyond the Anthropocene." *Religions* 6 (2015): 742–754.
Bauman, Whitney and Kevin O'Brien. Forthcoming Manuscript. *An Environmental Ethics of Uncertainty: The Pace of Ambiguity in a Time of Unambiguous Trouble.*
Bennett, Jane. *Vibrant Matter: A Political Ecology of Things*. Durham, NC: Duke University Press, 2010.
Brennan, Teresa. *Globalization and Its Terrors: Daily Life in the West*. New York: Routledge, 2003.
Butler, Judith. *Bodies That Matter: On the Discursive Limits of Sex*. New York: Routledge, 1993.
Carson, Rachel. *Under the Sea Wind*. New York: Simon & Schuster, 1941.
Davis, Jack. *An Everglades Providence: Marjory Stoneman Douglas and the American Environmental Century*. Athens, GA: University of Georgia Press, 2009.
Deleuze, Gilles and Felix Guattari. *A Thousand Plateaus: Capitalism and Schizophrenia*. Minneapolis, MN: University of Minnesota Press, 1987.
Derrida, Jacques. *Of Grammatology*. Baltimore, MD: Johns Hopkins University Press, 1974.
Halberstam, Judith. *The Queer Art of Failure*. Durham, NC: Duke University Press, 2011.
Haraway, Donna. *Staying with the Trouble: Making Kin in the Chthulucene*. Durham, NC: Duke University Press, 2016.
———. *Simians, Cyborgs and Women*. New York: Routledge, 1991.
———. "Situated Knowledges: The Science Question in Feminism and the Privilege of Partial Perspective." *Feminist Studies* 14 no.3 (1988): 575–599.

Harding, Sandra. *Is Science Multicultural? Postcolonialisms, Feminisms and Epistemologies*. Bloomington, IN: Indiana University Press, 1998.

Keller, Catherine. *Cloud of the Impossible: Negative Theology and Planetary Entanglement*. New York: Columbia University Press, 2014.

———. *Face of the Deep: A Theology of Becoming*. New York: Routledge, 2002.

Keller, Catherine and Laurel Schneider, eds. *Polydoxy: A Theology of Multiplicity and Relation*. New York: Routledge, 2010.

Latour, Bruno. *Politics of Nature: How to Bring the Sciences into Democracy*. Cambridge MA: Harvard University Press, 2004.

———. *We Have Never Been Modern*. Cambridge, MA: Harvard University Press, 1993.

Meadows, Donella H. and Jorgan Randers. *The Limits to Growth: A Report for the Club of Rome's Project on the Predicament of Mankind*. New York: Universe Books, 1972.

Morton, Nelle. *The Journey Is Home*. Boston, MA: Beacon, 1985.

Sideris, Lisa and Kathleen Dean Moore, eds. *Rachel Carson: Legacy and Challenge*. Albany, NY: State University of New York Press, 2008.

Stenmark, Lisa. *Religion, Science and Democracy: A Disputational Friendship*. Lanham, MD: Lexington, 2013.

———. "An Ecology of Knowledge: Feminism, Ecology and the Science and Religion Discourse," in *Metaviews* 6 (February 5, 2001). From the Metaviews Listserv: metaviews@meta-list.org.

Thompson, Paul B. *The Spirit of the Soil*. New York: Routledge, 1995.

White, Lynn. "The Historical Roots of Our Ecological Crisis." *Science* 155–3767(1967): 1203–1207.

Whitehead, Alfred North. *Process and Reality*. New York: The Free Press, 1978.

Žižek, Slavoj. *In Defense of Lost Causes*. London, UK: Verso, 2009.

NOTES

1. Catherine Keller, *Cloud of the Impossible: Negative Theology and Planetary Entanglement* (New York, NY: Columbia University Press, 2014).

2. Karen Barad, *Meeting the Universe Halfway: Quantum Physics and the Entanglement of Matter and Meaning* (Durham, NC: Duke University Press, 2007).

3. Judith (Jack) Halberstam, *The Queer Art of Failure* (Durham, NC: Duke University Press, 2011).

4. Donna Haraway, *Simians, Cyborgs and Women* (New York, NY: Routledge, 1991), 193.

5. Barad, *Meeting the Universe*, 132–188.

6. Karen Barad, "Nature's Queer Performativity," in *Kvinder, Køn & Forskning* 12 (2012): 32.

7. Ibid., 38.

8. Keller, *Cloud of the Impossible.*

9. Halberstam, *The Queer Art of Failure.*

10. Gilles Deleuze and Felix Guattari, *A Thousand Plateaus: Capitalism and Schizophrenia* (Minneapolis, MN: University of Minnesota Press, 1987), 310–350.

11. Catherine Keller and Laurel Schneider, eds., *Polydoxy: A Theology of Multiplicity and Relation* (New York, NY: Routledge, 2010).

12. Deleuze and Guattari would talk about this as the "virtual being the real" (Deleuze and Guattari, *A Thousand Plateaus,* 92–98); whereas Whitehead might simply say that the process is what is really real and individual entities (or moments) are abstractions from this process. See, e.g.: Alfred North Whitehead, *Process and Reality* (New York, NY: The Free Press, 1978), 73.

13. On abjection see: Judith Butler, *Bodies That Matter: On the Discursive Limits of Sex* (New York, NY: Routledge, 1993), xiii. On "remainders" and "traces" and "differance" see, e.g.: Jacques Derrida, *Of Grammatology* (Baltimore, MD: Johns Hopkins University Press, 1974), 65–79.

14. Donna Haraway, "Situated Knowledges: The Science Question in Feminism and the Privilege of Partial Perspective," *Feminist Studies* 14.3(1988): 575–599.

15. Sandra Harding, *Is Science Multicultural? Postcolonialisms, Feminisms, and Epistemologies* (Bloomington, IN: Indiana University Press, 1998), 124–145.

16. On "the collective" see: Bruno Latour, *Politics of Nature: How to Bring the Sciences into Democracy* (Cambridge, MA: Harvard University Press, 2004).

17. Lisa Stenmark, *Religion, Science and Democracy: A Disputational Friendship* (Lanham, MD: Lexington, 2013).

18. Whitehead, *Process and Reality*, 7.

19. Nelle Morton, *The Journey Is Home* (Boston, MA: Beacon, 1985), 206.

20. Barad, "Nature's Queer Performativity," 46.

21. For a nice introduction to the "New Materialism" see: Jane Bennett, *Vibrant Matter: A Political Ecology of Things* (Durham, NC: Duke University Press, 2010).

22. Judith Halberstam, *The Queer Art of Failure*, 3.

23. Paul B. Thompson, *The Spirit of the Soil* (New York, NY: Routledge, 1995).

24. Teresa Brennan, *Globalization and Its Terrors: Daily Life in the West* (New York, NY: Routledge, 2003).

25. I discuss at length my reasons for this in the article: Whitney Bauman, "Climate Weirding and Queering Nature: Getting Beyond the Anthropocene," *Religions* 6 (2015): 742–754.

26. Donna Haraway, *Staying with the Trouble: Making Kin in the Chthulucene* (Durham, NC: Duke University Press, 2016).

27. This is the infamous "Lynn White" Argument: "The historical roots of our ecological crisis," *Science* 155:3767 (1967):1203–1207.

28. Teresa Brennan, *Globalization and Its Terrors*, 138.

29. Catherine Keller, *Cloud of the Impossible: Negative Theology and Planetary Entanglement* (New York, NY: Columbia University Press, 2014), 214.

30. The text from the following sections on Douglas and Carson is adapted from my recent manuscript with Kevin O'Brien, *An Environmental Ethics of Uncertainty: The Pace of Ambiguity in a Time of Unambiguous Trouble* (currently under review).

31. Jack Davis, *An Everglades Providence: Marjory Stoneman Douglas and the American Environmental Century* (Athens, GA: University of Georgia Press, 2009).

32. Nelle Morton, *The Journey Is Home*, 206.

33. Davis, *An Everglades Providence*, 767.

34. Ibid., 809.

35. Ibid. 828.

36. Lisa Sideris and Kathleen Dean Moore, eds., *Rachel Carson: Legacy and Challenge* (Albany, NY: State University of New York, 2008), 7.

37. Ibid. 6.

38. Ibid.: see chapters 3 and 4.

39. Donella H. Meadows and Jorgan Randers, *The Limits to Growth: A Report for the Club of Rome's Project on the Predicament of Mankind* (New York: Universe Books, 1972).

40. Sideris and Moore, *Rachel Carson*, 8.

41. Lisa Stenmark, "An Ecology of Knowledge: Feminism, Ecology and the Science and Religion Discourse," in *Metaviews* 6 (February 5, 2001). From Metaviews Listserv: metaviews@meta-list.org.

42. Rachel Carson, *Under the Sea Wind* (New York, NY: Simon & Schuster, 1941), 109.

43. Catherine Keller, *Face of the Deep: A Theology of Becoming* (New York, NY: Routledge 2002).

Chapter Five

Polyamorous Bastards

James Baldwin and Desires of a Queer African-American Religious Naturalism

Carol Wayne White

At the height of the civil rights era in the twentieth century, James Baldwin poignantly described blacks' experiences of marginality in North America. In a country besieged by white supremacy, he tried to capture the acute sense of displacement felt by African Americans with the creative use of the "bastard" epithet. In doing so, he drew richly from formative familial experiences and the black holiness tradition of his youth. Having never known his biological father, and feeling estranged from his emotionally distant stepfather—a factory worker and storefront preacher whom his mother married when he was three—James Baldwin would later write about the anguish of experiencing life as an "illegitimate" kid. In both acclaimed novels and critical essays, he creatively used the bastard motif to augment a critical self-awareness associated with being the outsider and the illegitimate other. Conjoining the personal and the political, Baldwin rhetorically expanded the term bastard to convey blacks' harried existence in the "New World" and to evoke an ethical quandary for white Americans. In the absence of embodied authenticity and relational integrity, Baldwin's bastard metaphor revealed the pathology inherent in many whites' refusal to embrace their familial kinship with blacks. With this term, Baldwin also raised a critical question to the America of his day: whether hope for its future could possibly exist in light of distorted forms of relationality. Baldwin's very use of the bastard term symbolized the moral paralysis he saw embedded in an American psyche suffering from a great lie perpetuated by white supremacy.

Baldwin's use of the bastard metaphor has elicited a range of critical responses, from Cornel West's distrust of a perceived apolitical individualism to Clarence Hardy's declaration that Baldwin was poignantly seeking recognition from whites.[1] However, I consider Baldwin's use of it as part of a complex ontological project with monumental social, affective, and ethical implications. With the bastard motif, Baldwin also underscored impoverished views of humanity kept in place by polarized, binary constructions (e.g., white/black, insider/outsider, superiority/inferiority, the saved/damned, hetero-normative/homo-depraved beings). In an attempt to address these various forms of alienation experienced by Americans desiring profound connection with otherness, Baldwin introduced the concept of love. Far from being a lofty abstraction, Baldwin's conception of love entailed a radical re-adjustment of human relations, requiring individuals and communities to embrace others they often feared, dismissed, or even hated. This specific form of love demanded from whites and blacks unprecedented acts of courage and audacious choices. Accordingly, his view of love was a creative means of synthesizing the affective and ethical with socio-political advancements.

In this chapter, I explore the rich conceptual space opened by Baldwin's use of the bastard metaphor and his ensuing concept of love.[2] In describing Baldwin's usage of this term, I focus primarily on cultural critiques found in his critical essays; I also use Baldwin's writings as an opening for articulating my fuller explication of queer religious naturalism. Inspired by his creative reach, I thus specify bastard as a trope to mark the emergence of an African-American religious naturalism that resists normative (and, in my rendering, impoverished) views of our humanity. Specifically, in negating pauperized views of blacks' humanity perpetuated by white supremacy, this African-American religious naturalism invites contemporary readers to reconsider who and what we are: value-laden natural processes that become human in specific orientations. It presupposes human animals' deep, inextricable homology with each other, drawing our attention to an expansive view of our humanity as an emergent phenomenon, not an achievement. Further, in underscoring humans' inexhaustible connection or entanglement with other natural processes, this naturalistic view also brings to light our essential connection with the more-than-human that constitutes the very notion of the human as such.

Further, building on a notion of love that I glean from Baldwin's self-understanding as an (sexualized) outsider, my view of religiosity celebrates nomadic, polyamorous relations. With its naturalistic grounding, this model of religiosity resists the "isms" based on binary constructions that uphold asymmetrical relationships and polarize our desire for connection with all that is. In so doing, this African-American religious naturalism adopts a queer positionality—or what Michael Warner has described as resistance "to regimes of the normal."[3] As queer enactment, this African-American religiosity calls for a radical relationality in which our experiences of love overcome arbitrary boundaries held in place by normalizing cultural markers. It seeks a modality of existence based in transformation; in such a vision, our expanded humanity as sentient beings is porous—we suffuse each other with care and a sense of belonging together.

To advance my argument, I first introduce Baldwin's cultural critiques that gave rise to the bastard motif and discuss his creative use of the term in specific contexts. These brief discussions serve as an opening to my fuller explorations of the term in a queer religious context. I then highlight Baldwin's rejection of a model of Christianity that glorified a depraved blackness with its theistic symbolism of a white God. Of key importance here are Baldwin's warnings against investing in a religious vision that implicitly kept in place white racist constructs, problematic cultural practices, and heteronormative values. Following this, I discuss Baldwin's concept of embodied love, which he viewed as a corrective to the normative religious system. Finally, I explore the possibility of a queer African-American religious naturalism that advances Baldwin's notion of enacting boundless love with each other. Given the historical gap between Baldwin's and our own time, I attempt to sketch a capacious model of African-American religiosity that he was unable to conceive at that time. In order to distinguish this religious perspective as queer enactment, I incorporate insights from Claudia Schippert's strategy of queering the religious discipline. My goal is to draw the contours of a queer African-American religious topography where polyamorous bastards roam ecstatically in nomadic desiring. Before beginning, my analysis, I must note an important observation about Baldwin's critiques. Given the historical framework in which he was writing, his critiques targeted a problematic, limited binary constituted by the symbolic notions of whiteness and blackness. As current critical race theory suggests, however, racist discourse emerges from the dominance of a white supremacist ideology, or a master narrative, that has failed to include the value and experiences of all groups whose identities have been isolated from and seen as distinct from "white" skin and everything associated with this distinction. In my broader work on religious naturalism, I emphasize this broader notion of racialized discourse as pertaining to all marginalized voices and groups whose experiences have never been legitimized within this master narrative.

BALDWIN'S BASTARDS: ILLICIT RACE AND NOMADIC SENSIBILITIES

Baldwin emerged as an essayist in the fifties when the civil-rights movement was barely in discernible form. In his first collection of essays, *Notes of a Native Son* (1955), followed by *Nobody Knows My Name* (1961) and *The Fire Next Time* (1963), Baldwin introduced a style of writing that fused in a unique fashion the personal, the political, and the literary. Whether recounting his experiences as a teenage preacher, a queer male, a black expatriate, a public intellectual, or an activist in the struggle for civil rights, Baldwin wrote passionately about the quandaries of living life as an outlier. He brought an intense self-awareness to his inhabiting within one body many different (and often excluded) cultural markers. In the America of his day, these identities were persistently viewed as oppositional: his blackness and erotic-affective desires; his precocious religious insights and radical activism; his designation as a black American artist in Europe juxtaposed with his European celebrity in white America. In his quest to create an authentic mode of existence, however, Baldwin resisted, with varying degrees of success, the easy option of reducing his capacious humanity to any single identity. Furthermore, he creatively expanded this sensibility to include the collective experiences of African Americans as he sought new ways of being in which he and others could live fully with perceived differences.[4]

In their introduction to a critical collection of essays on Baldwin, Cora Kaplan and Bill Schwartz described a dominant trend in Baldwin studies that often failed to understand this point. As they observed: "For too long one Baldwin has been pitted against another Baldwin, producing a series of polarities that has skewed our understanding: his art against his politics, his fiction against his nonfiction; his early writings against his late writings; American Baldwin against European Baldwin; black Baldwin against queer Baldwin."[5] I share this robust reading of Baldwin. Select essays show Baldwin resisting facile, externally imposed views of his humanity, as well as that of other blacks. He opened an imaginative space in which North Americans entrapped in isolationist encampments could be free and re-envision themselves as relational beings capable of living with difference. In the 1984 introduction to *Notes of a Native Son*, Baldwin described the menacing effects of racialized living in America. Symbolically, socially, and materially, human lives were affected by a problematic cultural system that elevated certain bodies over others, ultimately leaving all alienated from each other. For Baldwin, racialized existence in its most extreme form was expressed in a black-white oppositional logic that was omnipresent: "The conundrum of color is the inheritance of every American, be he/she legally or actually Black or White. It is a fearful inheritance, for which untold multitudes, long ago, sold their birthright. Multitudes are doing so, until today."[6]

Baldwin's writings reverberate with the conundrum of affirming life and embracing one's humanity in a world (or culture of values) that one has not created. As such, they reveal an important observation about black lives in America that Frederick Douglass articulated in the nineteenth century when addressing the National Colored Convention of 1853: "Our white fellow-country men do not know us. They are strangers to our character, ignorant of our capacity, oblivious of our history and progress, and are misinformed as to the principles and ideas that control and guide us as a people. The great mass of American citizens estimate us as being a characterless and purposeless people."[7] Baldwin experienced these racial distortions of his (and by extension, other blacks') humanity in the United States while seeking a sense of authentic selfhood. In the beginning of "Nobody Knows My Name," he declared: "I left America because I doubted my ability to survive the fury of the color problem here [. . .] I wanted to prevent myself from becoming *merely* a Negro; or even, merely a Negro writer. I wanted to find out in what way the specialness of my experience could be made to connect me with other people instead of dividing me from them."[8] Here I imagine Baldwin as a scriptor of human possibilities, seeking an expansive view of the human capacity to relate and connect with others without being restricted by problematic racial constructs. With refreshing candor, he targeted white supremacy as a cultural value system that functioned to reduce and obscure his ability to connect with others (and with himself) on a more fundamental, existential level. In such a system, his blackness—its symbolic resonance and its tactile materiality—became a source of alienation and disconnection based upon the disconcerting white gaze.

As Baldwin also discovered, another form of this alienation would be felt beyond American borders in a purportedly enlightened Europe. In the well-known essay, "Stranger in the Village," originally published in 1953, Baldwin recounted his experiences living in a tiny Swiss village. Reflecting on what it meant to be a stranger thrown in a raced world that he had not created, Baldwin observed that the people of the village cannot be, from the point of view of power, strangers anywhere in the world. He wrote about the children's response to him as an exotic rarity as they shout *Neger! Neger!* in the streets, oblivious to his reaction.[9] Notwithstanding the *saluts* and *bonsoirs* that Baldwin exchanged with his neighbors under the social convention of politeness, he saw in their eyes elements of paranoia and malevolence. In these encounters, Baldwin understood (or grasped) his "black" identity in personal and passionate terms. Notably, for Baldwin, this encounter with self was not a result of a disembodied, detached abstraction, but a deep involvement in the concreteness and full materiality of life. His experience in the Swiss village led him to reflect on the European roots of white racism and its later appearance in the United States: "For this village brings home to me this fact: that there was a day, and not really a very distant day, when Americans were scarcely Americans at all but discontented Europeans, facing a great unconquered continent and strolling, say, into a marketplace and seeing black men for the first time."[10]

The visibility of blackness that Baldwin alluded to—the very sight of black skin—measures the importance of materiality in his racial discourse. This racialized discourse also provided the backdrop for Baldwin's bastard metaphor, alerting readers to his aspirations of acquiring ontological wholeness both for himself and others. Later, while reflecting on his sojourns in Europe, and implicitly conjoining the singular with the collective, Baldwin inscribed a type of existence for black Americans that could at best be described, pathetically, as non-essential and marginal. In "Autobiographical Notes" from *Notes of a Native Son*, Baldwin asserted:

> I know, in any case, that the most crucial time in my own development came when I was forced to recognize that I was a kind of bastard of the West; when I followed the line of my past I did not find myself in Europe but in Africa. And this meant that in some subtle way, in a really profound way, I brought to Shakespeare, Bach, Rembrandt, to the stones of Paris, to the cathedral at Chartres, and to the Empire State Building, a special attitude. These were not really my creations, they did not contain my history; I might search in them in vain forever for any reflection of myself. I was an interloper; this was not my heritage.[11]

The cultural resonance of Baldwin's bastard metaphor here is noteworthy. The cultural artifacts he encountered that purportedly symbolized the best of human aspirations, desires, hopes, and creativity did not reflect his contributions as an African American. With sobering awareness, his sense of illegitimacy is heightened when he realizes that in this Euro-American cultural lineage, he was essentially being confronted with prevailing configurations of the normative human: an ideology of whiteness. Moreover, for Baldwin, experiencing oneself as a bastard evoked a sense of not knowing one's true, fuller identity. Indeed, black bastards in the United States are perpetually hidden from their truest selves when forced daily to wear a cloak of shame or inferiority that has been woven by a cultural legacy built on white supremacy. Equally important, in Baldwin's observations, American whites selfishly and ingeniously denied their culpability in creating this legacy of differentiation; in re-inventing—in more and more ingenious ways—the omnipresence of whiteness, white Americans intensified the alienating effects of problematic racial constructions. As he wrote, " . . . the price the white American paid for his ticket was to become white -: and, in the main, nothing more than that, or, as he was to insist, nothing less."[12]

With an emphasis on materiality and corporeality, Baldwin asserted again and again that black and white Americans are blood relatives—they shared the same biological origins. What his contemporaries often perceived as a racial problem essentially masked a more fundamental problem: forgetfulness of our common humanity. As he declared in *No Name in the Street*: "The problem is rooted in the question of how one treats one's flesh and blood, especially one's children. The blacks are the despised and slaughtered children of the great

Western house—nameless and unnamable bastards."[13] With a critical awareness of experiencing oneself, one's people, one's culture as not quite genuine—as irregular, inferior, or of dubious origin—Baldwin spoke of cultivating a special attitude, "[a] special place in this scheme."[14] While not denying the reality of cultural, historical forces, Baldwin also declared that humans are always so much more than what our cultural markers claim for us. As he realized, . . . "I had to claim my birthright. I am what time, circumstance, history, have made of me, certainly, but I am, also, much more than that. So are we all."[15]

These passages help underscore the ontological implications implicit in Baldwin's critical discourse. Claiming one's heritage is part of a more complex process of actualizing oneself as one relates to others, aspiring to achieve and experience one's humanity without falling prey to the damaging effects of a binary system that demarcates some humans as more, others as less. As he noted, the problematic question of African Americans' "humanity," and of their "rights as human being[s]" became such a burning question for several generations of Americans that it ultimately "became one of those used to divide the nation."[16] However, for Baldwin, his—and all blacks'—heroic acts of self-actualization were not enough to address the great racial divide. Courageous acts by whites were crucial in helping to transform America; specifically, what was essentially needed was the honest admission by white Americans of their pivotal role in creating the current state of affairs in the America of his day: bastardizing blacks' existence there. In short, naming and forgoing the normativity of whiteness in America was tantamount. Historically, this meant whites' admission of the systemic, widespread cultural violence perpetuated by the state in its construction and adoration of whiteness. White Americans must acknowledge, first and foremost, this shared history for blacks and whites. In *The Price of the Ticket*, Baldwin states:

> The record is there for all to read. It resounds all over the world. It might well be written in the sky. One wishes that Americans—white Americans—would read, for their own sakes, this record and stop defending themselves against it. Only then will they be enabled to change their lives. The fact that they have not yet been able to do this—to face their history, to change their lives—hideously menaces this country. Indeed, it menaces the entire world.[17]

According to Baldwin, the subjugated history perpetuated by white Americans was the key to their identity, as well as the triumph and justification of their history; furthermore, their material well being depended on this continued subjugation. He asserts:

> One may now see that the history, which is now indivisible from oneself, has been full of errors and excesses, but this is not the same thing as seeing that, for

> millions of people, this history—oneself—has been nothing but an intolerable yoke, a stinking prison, a shrieking grace. It is not so easy to see that, for millions of people, life itself depends on the speediest possible demolition of this history, even if this means the leveling, or the destruction of its heirs.[18]

To be heirs of history, in this context, is to construct and maintain the reality that one desires. Outsiders, or bastards, Baldwin suggests, are denied this very act of creating reality for oneself. In *No Name in the Street* (1972), Baldwin asserted that white supremacy operates historically in prohibiting blacks in America (and the "black" Algerians' in France) from constructing their own freedom and choosing to become what they may desire. In other words, their experiences within the subjugated history of whiteness amounts to an essential lack of access to oneself: "The Algerian and I were both, alike, victims of this history, and I was still a part of Africa, even though I had been carried out of it nearly four hundred years before."[19] The radical nature of materiality and historical positioning for Baldwin are notable here as he resists the pernicious violence perpetrated by history in order to affirm differently unrecognized—or bastards'—accounts of history. Thus, an important insight of Baldwin, which I will emphasize in promoting my model of religious naturalism, is that humans do not transcend our variegated identities in the act of affirming our common humanity. Rather, we affirm our identities through a radically refigured conception of our humanity, which always includes historical specificity, diversity, and dynamic processes of becoming. In other contexts, Baldwin also alluded to the formation of a bastard (or outsider's) status that keeps blacks intimately identified with exile—or of not knowing home in the way that white Americans have created home in the United States—and thus of never being accepted, fully known, or even embraced by their white counterparts. He describes the tragic sense of forlornness that black Americans feel in experiencing a diaspora that never ends. As he describes it, in the midnight hour when one's "missing identity aches," and when one "can neither access nor overcome the storm of the middle passage. One is mysteriously shipwrecked forever, in the Great New World."[20]

BLACK BASTARDS, DESIRE, AND RELIGIOUS ENTRAPMENTS

Baldwin's vision of a nobler view of humanity led him to indict the construction of whiteness in the United States, which both aided in demarcating and separating a common humanity that all shared, and fostered myriad forms of alienation and hatred among various individuals and groups. In dismantling this edifice, he targeted a major cultural institution: religion, and specifically Christianity. Baldwin keenly recognized that key religious ideas functioned (either explicitly or implicitly) in a racist culture essentially to devalue black bodies as unworthy and inherently inferior to white ones, and they generated deeply embedded black self-loathing among many African Americans. In *To Crush the Serpent* (1987), one of his final published essays, Baldwin summed up a theme that he had addressed throughout many earlier ones: "Race and religion, it has been remarked, are fearfully entangled in the guts of this nation, so profoundly that to speak of the one is to conjure up the other. One cannot speak of sin without referring to blackness, and blackness stalks our history and our streets."[21]

Baldwin addressed the wider matrix of cultural meanings inhering in religious systems, showing how they both shape, and are shaped, by the behaviors of many black and white U.S. citizens. Within this context, the bastard metaphor was illuminating, as it helped Baldwin to underscore a truth that most wanted to ignore: white Americans' denial of any familial kinship with blacks. As Baldwin observed in the interview with James Mossman, "The great dilemma of being a white American precisely is that they deny their only kinship."[22] He specifically targeted the hypocrisy of white Christians who, in an attempt to maintain their social order identity as white people, willfully denied their own moral connection (and biological kinship) to those of African descent, whom they continued to exploit.

In both dominant white Christian culture and the holiness tradition of his youth, which he saw as absorbing problematic ideological aspects of the former, Baldwin identified a root problem, manifest in various ways and on different levels: systematic vilification of blackness. He specifically targeted the symbol God as a reified marker that posited whiteness as representative of aesthetic and moral truths; equally important, it established whiteness as constitutive of normative humanity. Furthermore, whiteness was a social identity rooted in a god-complex, representing dominant culture's desperate avoidance of its own limitations as whites denied the beauty, value, and complex humanity of African Americans. He wrote at length about the depth of anti-blackness pervasive in American culture subtlety embedded in religious mechanisms. In *The Fire Next Time*, Baldwin observed:

> Negroes in this country—and Negroes do not, strictly or legally speaking, exist in any other—are taught really to despise themselves from the moment their eyes open on the world. This world is white and they are black. White people hold the power, which means that they are superior to blacks (intrinsically, that is: God decreed it so), and the world has innumerable ways of making this difference known and felt and feared.[23]

In linking the notion of white supremacy with theistic belief and doctrinal certainty, Baldwin revealed why some cultural values entrenched in racial distortions appear as real, enduring, and authoritative truths. In the face of unrelenting racism and brutality, Baldwin thus saw structured religion as an obstacle to blacks' achieving authentic selfhood and integrity of being. In *To Crush the Serpent*, for example, Baldwin discussed the adverse psychological effects of those white Christian theological systems that invoked the curse of Ham to both justify slavery and devalue black subjectivities.[24] The rabid anti-blackness in such dangerous fabrications was alarming to Baldwin, who unmasked them as distortions of empirical truths.

Even more troubling for Baldwin were the insidious psychological effects on blacks of paying homage to the Christian deity. In commenting on the paradoxical nature of worship in the God-intoxicated holiness tradition of his youth, he observed: "But God—and I felt this even then, so long ago, on that tremendous floor, unwilling—is white. And if His love was so great, and if He loved all His children, why were we, the black, cast down so far?"[25] Although he acknowledged the black church (in all of its various structures) was often a haven and site of communal safety from blatant acts of racism, Baldwin also saw its complicity with white supremacy as blacks harbored fear and hatred of their own bodies and those that looked like theirs. He perceived a form of self-hatred embedded in his tradition's ritualistic fervor and rites of purification where, ironically, distraught religious adherents often denied themselves the healing, love, and pleasure they were entitled to experience: "And the passion with which we loved the Lord was a measure of how deeply we feared and distrusted, and, in the end, hated almost all strangers, always, and avoided and despised ourselves."[26]

This awareness led Baldwin to another troubling insight: one unfortunate outcome of internalized anti-blackness was a lack of love in the black Church. In *The Fire Next Time,* Baldwin mused, "I really mean that there was no love in the church," adding that it was a mask for "hatred and self-hatred and despair."[27] Baldwin experienced and witnessed the degree to which this self-hatred—extraordinarily moralistic in tone—was a guise for advancing anti-body, anti-sexual, xenophobic, and homophobic sentiments. Such anti-love fostered a passivity and repression of everything that he and others experienced as naturally good, and it required the relinquishing of one's individuality. Amid the violent forms of alienation caused by problematic racial distinctions, Baldwin introduced a radical view of love. For Baldwin, love is a term that describes a state of being one affirms again and again in the process of choosing to enact one's authentic humanity:

> Love takes off the masks that we fear we cannot live without and know we cannot live within. I use the word "love" here not merely in the personal sense

but as a state of being, or a state of grace—not in the infantile American sense of being made happy but in the tough universal sense of quest and daring and growth.[28]

With his usual flair and emotional intensity, Baldwin conjoined the private and the public, the personal and the political, to describe a transformative embodied love that brings awareness of our common humanity. In "No Name in the Street," his eloquence is breathtaking when describing falling in love with another concrete, material human being—an experience that unhinges the entrapments of racial constructs:

> It began to pry open for me the trap of color: for people do not fall in love according to their color—this may come as news to noble pioneers and eloquent astronauts, to say nothing of most of the representatives of most of the American states—and when lovers quarrel, as indeed they inevitably do, it is not the degree of their pigmentation that they are quarreling about, nor can lovers, on any level whatever, use color as a weapon. This means that one must accept one's nakedness. And nakedness has no color: this can come as news only to those who have never covered, or been covered by, another naked human being.[29]

Baldwin's embodied love described above evokes the Sartrean notion of intersubjectivity: recognizing the humanity of the other before oneself confronts one's subjectivity in the most immediate way, both limiting and enabling what one could possibly choose in any given context. This intimate encounter of knowing and being known by another brings with it a new awareness of seeing others differently and experiencing one's humanity as both free and bound. As Baldwin stated, an individual is both stronger and more vulnerable, both free and bound: "Free, paradoxically, because, now, you have a home—your lover's arms. And bound: to that mystery, precisely, a bondage which liberates you into something of the glory and suffering of the world."[30]

The embodied love Baldwin evoked also imbued traditional religious terms with new fresh, expanded meanings. His radical view of love was "something active, more like a fire, like a wind," not an empty abstraction describing a passive stance before some authorial figure outside of oneself.[31] Here, we see Baldwin rejecting the traditional otherworldly eschatological discourse of fear and damnation featured in holiness traditions, and replacing it with an emphasis on the concrete dynamics of living here and now. Likewise, salvation is that which we must do to save each other; for Baldwin, the most crucial aspect of salvation is its rootedness in human actions and efforts. The contingencies of life and concreteness of human experiences require redemptive actions from humans themselves. In "To Crush the Serpent" Baldwin asserted that salvation does not divide, but conjoins, so that " . . . one sees oneself in others and others in oneself. It is not the exclusive property of any dogma, creed, or church. It keeps the channel open between oneself and however one wishes to name that which is greater than oneself."[32]

Baldwin believed embracing embodied love could result in a vital flourishing for all North Americans. He reiterated this theme in a speech he gave in San Francisco in October of 1960. Addressing the writer's role in American life, Baldwin emphasized and articulated a moral vision that celebrates the potential of newly formed human relationships to create and sustain new possibilities for Americans. In his thinking, humans displace the traditional God and enact transformation in their lives, redeeming themselves from impoverished, erroneous views of their shared humanity. For example, after insisting on necessary changes to the dominant configuration of raced living in America, Baldwin ended the speech with the following words about the nation: "It will not be transformed by an act of God, but by all of us, by you and me. I don't believe any longer that we can afford to say that it is entirely out of our hands. We made the world we're living in and we have to make it over."[33]

In Baldwin's hands, traditional African-American religiosity is destabilized and uprooted from a metaphysical system that pits a supernatural deity over against sinful humans. He displaced traditional supernaturalism, exchanging the external deity beyond nature for the power of love expressed in embodied, material human relationships. In short, for Baldwin, humans save each other. As D. Quentin Miller suggested, one can see Baldwin moving from the ultimate expression of external authority—God—to the broader community collectively and individually.[34] Furthermore, the fuller expression of Baldwin's form of communal ontology recognized a common humanity constitutive of our biotic materiality on which various identity markers are couched and binary oppositions attached. As he emphatically stated at one point: "It is so simple a fact and one that is so hard, apparently, to grasp: *Whoever debases others is debasing himself.*"[35] Here, Baldwin anticipated a fundamental theme of my religious naturalism, which emphasizes the deep genetic homology structuring all life forms—what I describe as humans' interconnectedness with each other and with all natural organisms.

OUR COMMON HUMANITY: EMERGENCE OF AN AFRICAN-AMERICAN RELIGIOUS NATURALISM

An African-American religious naturalism emerges out of a critical awareness that religiosity is not necessarily centered in any specific tradition; rather, it can be a mode of reflecting on, experiencing, and envisioning one's relationality with all that is. I evoke the views of Peter Van Ness, who wrote eloquently of the spiritual dimension of life "as the embodied task of realizing one's truest self in the context of reality apprehended as a cosmic totality. It is the quest for attaining an optimal relationship between what one truly is and everything that is."[36] Since religious naturalism does not use supernatural concepts or theories in comprehending humans' need for value and meaning, the realm of nature is the focus (this includes both natural processes and human culture). Within the context of African-American life and culture, this means that any truths we are ever going to discover, and any meaning in life we will uncover, are revealed to us through our own efforts as natural beings. Consequently, a fuller emergence of this African-American religious naturalism is possible if, and only if, we continue to keep our focus on artful, material human organisms, or on the efforts of relational humans.

Religious naturalists draw on two fundamental convictions in understanding basic human quests for meaning and value: the sense of nature's richness, spectacular complexity, and fertility, and the recognition that nature is the only realm in which people live out their lives. Donald Crosby provided an elegant summary of the prominent status of nature in religious naturalism:

> Nature requires no explanation beyond itself. It always has existed and always will exist in some shape or form. Its constituents, principles, laws, and relations are the sole reality. This reality takes on new traits and possibilities as it evolves inexorably through time. Human beings are integral parts of nature, and they are natural beings through and through. They, like all living beings, are outcomes of biological evolution. They are embodied beings whose mental or spiritual aspect is not something separate from their bodies but a function of their bodily nature. There is no realm of the supernatural and no supernatural being or beings residing in such a realm.[37]

Nature itself becomes a focal point for assessing our human desires, dreams, and possibilities—for assessing what can emerge from the past. Fifty years removed from Baldwin's nomadic sensibilities and his emphasis on materiality and embodied desire, I envision a religious naturalism that requires us to take seriously the idea of our humanity as an achievement, not a given. More specifically, this emergent religious naturalism compels African Americans to reflect meaningfully on the emergence of matter (and especially life) from the Big Bang forward, promoting an understanding of myriad nature as complex processes of becoming. Its theoretical appeal is the fundamental conception of humans as natural processes intrinsically connected to other natural processes. This insight helps to blur the arbitrary ontological lines that human animals have erected between other species and natural processes and us.

With Loyal Rue, I endorse a portrayal of human beings as star-born, earth-formed creatures endowed by evolutionary processes to seek reproductive fitness under the guidance of biological, psychological, and cultural systems that have been selected for their utility in mediating adaptive behaviors.[38] Humans maximize their chances for reproductive fitness by managing the complexity of these systems in ways that are conducive to the simultaneous achievement of personal wholeness and social coherence. Rue wrote:

> The meaning of human life should be expressed in terms of how our particular species pursues the ultimate telos of reproductive fitness. Like every other species, we seek the ultimate biological goal according to our peculiar nature. That is, by pursuing the many teloi that are internal to our behavior mediation systems, whether these teloi are built into the system by genetic means or incorporated into them by symbolic means. For humans there are many immediate teloi, including the biological goals inherent in drive systems, the psychological goals implicit in our emotional and cognitive system, and the social goals we imbibe through our symbolic systems. Human life is about whatever these goals are about.[39]

Appreciating human life as one distinct biotic form emerging from, and participating in, a series of evolutionary processes that constitute the diversity of life has monumental implications for African-American culture. Here, the scientific epic becomes *the* starting point for positing an African-American religiosity constituted by a central tenet: humans are relational processes of nature; in short, we are nature made aware of itself. In declaring such, I contend that our *humanity* is not a given, but rather an achievement.

Consider that from a strictly biological perspective, humans are organisms that have slowly evolved by a process of natural selection from earlier primates. From one generation to another, the species that is alive now has gradually adapted to changing environments so that it could continue to survive. Our animality, from this perspective, is living under the influence of genes, instincts, and emotions, with the prime directive to survive and procreate. Yet, this minimalist approach fails to consider what a few cognitive scientists, and most philosophers, humanists, and religionists tend to accentuate: our own personal experience of what it is like to be an experiencing human being. Becoming human, or actualizing ourselves as human beings, in this sense, emerges out of an awareness and desire to be more than a conglomeration of pulsating cells. It is suggesting that our humanity is not reducible to organizational patterns or processes dominated by brain structures; nor do DNA, diet, behavior, and the environment solely structure it. Human animals become human destinies when we posit fundamental questions of value, meaning, and purpose to our existence. Our coming to be human destinies is structured by a crucial question: How do we come to terms with life?[40]

In this naturalistic religious view, sacrality is a specific affirmation and appreciation of that which is fundamentally important in life, or that which is ultimately valued: relational Nature. Humans are interconnected parts of Nature, and our sacrality is a given part of Nature's richness, spectacular complexity, and beauty. Notwithstanding the diverse cultural and individual approaches of articulating this truth, there is for me, quite simply, the sacrality of human deep interconnectedness with all that is. Finding meaning and value in our lives within the natural order presupposes our fundamental interconnectedness. We can claim and become our humanity in seeking and finding community with others—and with otherness. This is a simple value that religious discourse has advanced and reiterated again and again. As Ursula Goodenough has observed:

> We have throughout the ages sought connection with higher powers in the sky or beneath the earth, or with ancestors in some other realm. We have also sought, and found, religious fellowship with one another. And now we realize that we are connected to all creatures. Not just in food chains or ecological equilibria. We share a common ancestor. We share genes for receptors and cell cycles and signal-transduction cascades. We share evolutionary constraints and possibilities. We are connected all the way down.[41]

The basic conception of the human as an emergent, interconnected life form amid spectacular biotic diversity has far-reaching ethical implications within the context of African-American culture. First, it contributes to an intellectual legacy that has attempted to overcome the deficient conceptions of our humanity couched in problematic racial constructions. African American religious naturalism presupposes human beings as biotic forms emerging from evolutionary processes sharing a deep homology with other sentient beings and also valuing such connection. Accordingly, it challenges racially constructed views that have persistently placed blacks outside of the circle of humanity.

Second, this model of African-American religiosity rejects a view of our humanity solely as an individualistic phenomenon—some type of communal ontology is implied. A crucial lesson here is that notwithstanding the cultural and national differences and specificities we construct, humans are all genetically connected and part of a greater whole—any harm done to another human is essentially harm done to ourselves. We are essentially celebrating a relational self that can resist solipsistic tendencies and egoistic impulses: there is no isolated self who stands over against the fields of interaction. Put another way, there is no private self or final line between interiority and exteriority—we always include the other (even if by acting to exclude it). The self is constitutionally relational and inevitably entangled in temporal becoming.

Finally, this African-American religious naturalism provides the impetus and vision for continued social justice action in the twenty-first century. With its emphasis on deep connectivity—with oneself, family, larger human community, local and global ecosystems, and the universe, this religious naturalism seeks a transformed existence. Religiously, this implies love, and love implies concern for the well being of the beloved. African-American religious naturalism reinforces perennial, expansive perspectives from the wisdom traditions that adamantly promote kindness, empathy, and compassion for all natural processes, including human ones. With the capacity to influence each other and other natural processes, humans have a responsibility to act in ways that promote the flourishing of all life, and to urge other humans that may be less inclined of our interconnectedness to do the same. Any inkling of white supremacy, or sense of cultural superiority of any ilk, is antithetical to this natural view; these eschewed cultural constructions are forced impositions on the wholeness of natural interrelatedness and deep homology that evolution has wrought.

BALDWIN'S AMERICA: FROM BASTARDS TO FREAKS

With its conception of human animals as value-laden organisms, African-American religious naturalism encourages us to challenge the most viral constructions of "isms" rooted in problematic self-other differentiations. Inspired by Baldwin's concept of embodied love, this religiosity resists normalizing discourses (often disguised as moral truths) that thwart human desire to connect with, and love, others. In short, it engages in queer resistance.

Because Baldwin did not fit standard heterosexual norms, people sought to label him. As he imagined a space to live authentically, Baldwin found certain categories reductive and impoverishing, failing to capture the full experience of human connection he sought. His queer sensibilities recognized that these labels were not merely handy descriptions in a culture that he had not helped to design; rather, they conveyed implicit moral and ontological meanings, and they were used to restrict his humanity and distort his nomadic desiring. During an interview with journalist James Mossman and Colin MacInnes in 1965, Baldwin asserted: "Those terms, homosexual, bisexual, heterosexual are 20th-century terms which, for me, really have very little meaning. I've never, myself, in watching myself and watching other people, in watching life, been able to discern exactly where the barriers were. Life being what life is, and passion being what passion is."[42] In keeping with my robust reading of Baldwin, I see this response as an enactment of queer resistance. Baldwin refused to see his desire to connect intimately with another male as an example of a "homosexual" summation. He recognized that categories of sexual identity are often used to demarcate and evaluate the wide range of erotic desire along a continuum, implying a range from the normative to the illegitimate. Moreover, Baldwin saw that "distrust of the affections and of the flesh . . . [is] revealed most grotesquely in what we call the sexual deviants, the sexual minorities, who are really simply the most vivid victims of our system of mortification of the flesh."[43]

In a 1985 essay titled "Here Be Dragons" (also published as "Freaks and the American Ideal of Manhood"), Baldwin also made fruitful connections among race, erotic-desiring beings, and nationalistic mythology. He observed that expressions of sexuality in the United States were rooted in an American ideal of masculinity, which perpetuate problematic binary differentiations: ". . . good guys and bad guys, punks and studs, tough guys and softies, butch and faggot, black and white."[44] Drawing on his personal experiences as a young black male exploring desire in Greenwich Village, Baldwin described lonely, alienating encounters with other males (often white or of another ethnicity) who both expressed desire for him as well as self-revulsion in desiring a black queer male. He also noted the virtual absence of black women in the Village when describing a few sexual encounters with white women whom he feared brought to their sexual encounters the desire to "civilize you into becoming their appendage," or with the individual one who desired "a black boy to sleep with because she wanted to humiliate her parents."[45]

In Baldwin's nomenclature of desire, he became a freak. Here, a key insight of his analysis is that freaks represent an epistemic marker that conjoins race, gender, and sexuality in U.S. culture. Freaks are deviant individuals who do not conform to the ideal standards of sexuality and normative configurations of masculine identity within a racist, homophobic culture. Baldwin observed: "Freaks are called freaks and are treated as they are treated—in the main, abominably—because they are human beings who cause to echo, deep within us, our most profound terrors and desires."[46] Accordingly, freaks uncloak a horrific and terrifying truth about purportedly "normal" people, providing insight into the dangers, violence, and weaknesses endemic to the American ideal of masculinity. As such, freaks remain outliers, sharing with bastards an illegitimacy that keeps the cultural myth intact.

As a site of memory, Baldwin's case is illuminative. Earlier he had critiqued the dangers of establishing the pure white god as the ontological grounding for an inferior humanity in which blacks represented the most depraved case. Here, in a parallel move, we see Baldwin exposing another sacrosanct truth based on whiteness: an ontological ordering that established white male heterosexuality the norm for ascertaining and evaluating other expressions of desire. Confronting this standard, Baldwin exposed a national psyche enshrouded in a fear of materiality, femininity, and blackness. For Baldwin, black gay, effeminate men are archetypes of this triple threat—they fascinate as phantasms and yet are reviled as actual humans. While limited in its lack of serious attention to female bodies, or lesbian desires, or even complex configurations of femininity, Baldwin's insights are nonetheless helpful, anticipating later queer interventions and orientations. Going beyond a vision of our humanity dictated by compulsory heterosexuality, he ends this absorbing essay with an image of an androgynous humanity where inter-related, inter-connected sexual, racial and gendered beings belong to each other. As he declared: "each of us, helplessly and forever, contains the other—male in female, female in male, white in black, and black in white. We are a part of each other."[47]

POLYAMOROUS BASTARDS: BECOMING OUR HUMANITY

Inspired by Baldwin's creative reach, my trajectory of African-American religious naturalism promotes a black queer critical consciousness that resists regimes of the normal. In this final section, I elaborate on this point, utilizing Claudia Schippert's work in queer theory and religion to underscore African-American religious naturalism as an instance of queer religiosity. In "Implications of Queer Theory for the Study of Religion and Gender: Entering the Third Decade," Schippert observed that queer theoretical analysis itself is an emergent discourse that is continually shifted and translated in diverse geographic-cultural configurations.[48] Furthermore, within the context of religious scholarship and studies, queer theory continues to expand its theoretical applicability and usefulness. For Schippert, then, queering religion has something to do with expanding the sphere and proper object of study in religion. This is what Schippert suggested with the term :undisciplined" religion:

> When approaching religion queerly then, we can no longer simply "add" queer identities to an "inclusive" liberationist agenda. . . . The implications of the challenges and rethinking of the field of religion indicate that what is being studied as the discipline of religion is shifting focus on previously excluded topics. Studying religion queerly can be, and perhaps will need to be, (more) undisciplined.[49]

The emergent African-American religious naturalism that I have sketched in this essay contributes to this undisciplined space. As a trajectory of religiosity in African-American culture, it destabilizes traditional religious methods that purport to establish humanity's desires of actualization on metaphysical views of a superior deity; this religiosity also problematizes the notion that religious ideas and ethical practices are necessarily grounded in transcendental ideals. Rather, it honors the complex ways value-laden human animals attempt to enact our desire for others, fulfilling emotional, physical, and psychological needs. Rejecting traditional onto-theological claims of reality as the source of our being, this religious perspective celebrates value-laden organisms fundamentally coming to terms with life, or making sense of their existence, in relationship with others. Compelled by the evolutionary epic, this queer African-American religious naturalism offers an expanded view of humans as intrinsically relational, natural organisms and vital centers of value—or humans as lovers of life, motivated by a desire for goodness and connection with all that is.

With an expansive view of our common humanity, a queer African-American religious perspective compels us to keep in focus the desires of marginalized, often forgotten humans—polyamorous bastards—who resist substantive ontologies and onto-theological loyalties. Inspired by Baldwin, Audre Lorde, Bayard Rustin, and countless other nameless bastards and freaks who dared to humanize their desire, it accentuates the pleasurable "roaming" of a queer love that celebrates humans' radical relationality with each other, and, I emphasize, with other natural processes. However, unlike the lonely repressed males sketched by Baldwin in "Here Be Freaks," polyamorous bastards wander with bold audacity, cherishing all forms of materiality. Here, we are careful not to erect new norms and classifications that further vilify perceivable deviant "others," specifically in gendered, racialized, class-marked, or sexual terms. Moreover, as I have argued elsewhere, any conceivable notion of our common humanity will be ontologically enmeshed and entangled with other forms of natural life.[50] In keeping with the tenets of religious naturalism, we consistently resist the aims of traditional humanism that have posited humans outside of myriad nature and eclipsed the interrelatedness of all natural processes.

Those of us inspired by this queer African-American religious naturalism recognize the inclination in humans to institutionalize our reified constructions as normative views of truth or standards of practice. Whatever principles we deem sacred or effective in shaping our desires can never be divorced from cultural constructions, social/political contexts, and institutionalizing processes. However, in seeking to reorder "issues concerning identities, nationalism, communities, and material practices," we remain open to newer forms of becoming.[51] In short, achieving our humanity remains an ongoing, critical task—we never arrive. The queer religiosity I outline here accentuates this point.

Finally, this emerging African-American religious naturalism evokes Baldwin's concept of redemptive embodied love, which he introduced to inspire and challenge his contemporaries to conceive and hope beyond what seemed immediate and obvious. In the contemporary era, queer religious practice finds all Americans confronting and rejecting ideologies that perpetually bifurcate our humanity, instilling a shameful sense of inferiority and unworthiness in some and a false sense of superiority in others. Focusing upon the variegated beauty that evolution has wrought, we celebrate our differences even while honoring our common humanity. Equally important, we unhinge problematic stigmas of unnaturalness, illegitimacy, or even freakishness from our engagement with others as we explore nomadic human desire. Here, Baldwin's prophetic words in *The Price of the Ticket* of reexamining everything we have been taught are apropos: "Go back to where you started, or as far back as you can, examine all of it, travel your road again and tell the truth about it. Sing or shout or testify or keep it to yourself: but *know whence you came*."[52]

WORKS CITED

Baldwin, James. "Introduction." In *The Price of the Ticket: Collected Non-Fiction, 1948–1985*, iv–xx. New York: St. Martin's Press, 1985.

———. "No Name in the Street." In *The Price of the Ticket: Collected Non-Fiction, 1948–1985,* 449–55. New York: St. Martin's Press, 1985.

———. *Baldwin: Collected Essays.* Edited by Toni Morrison. New York: The Library of America, 1998.

———. "Notes for a Hypothetical Novel." In *Nobody Knows My Name,* 141–152. New York: Vintage, 1992.

———. *The Cross of Redemption: Uncollected Writings.* New York: Vintage, 2010.

Baldwin, James, Fred Standley, and Louis H. Pratt. *Conversations with James Baldwin.* Jackson: University Press of Mississippi, 1989.

Crosby, Donald. *Living with Ambiguity: Religious Naturalism and the Menace of Evil*. Albany: State University of New York Press, 2008.

Douglass, Frederick. *Frederick Douglass: Selected Speeches and Writings.* The Library of Black America Series. Edited by Philip S. Foner and Yuval Taylor. Chicago: Chicago Review Press, 2000.

Goodenough, Ursula. *The Sacred Depths of Nature.* New York: Oxford University Press, 1998.

Hardy III, Clarence E. *James Baldwin's God: Hope, and Crisis in Black Holiness Culture* Knoxville: The University of Tennessee Press, 2003.

Kaplan, Cora and Bill Schwartz. *James Baldwin: America and Beyond.* Ann Arbor: The University of Michigan Press, 2011.

Miller, D. Quentin, ed. *Re-viewing James Baldwin: Things Not Seen.* Philadelphia: Temple University Press, 2000.

Rue, Loyal. *Religion Is Not about God: How Spiritual Traditions Nurture Our Biological Nature and What to Expect When They Fail*. Piscataway, N.J.: Rutgers University Press, 2005.

Schippert, Claudia. “Implications of Queer Theory for the Study of Religion and Gender: Entering the Third Decade.” *Religion and Gender* 1, no. 1(2011): 66–84.
Van Ness, Peter H. *Spirituality and the Secular Quest.* New York: Crossroad, 1996.
Warner, Michael, ed. *Fear of a Queer Planet: Queer Politics and Social Theory*. Minneapolis: University of Minnesota Press, 1993.
West, Cornel. *Prophesy Deliverance! An Afro-American Revolutionary Christianity*. Philadelphia: Westminster Press, 1982.
White, Carol Wayne. *Black Lives and Sacred Humanity: Toward an African American Religious Naturalism*. New York: Fordham University Press, 2016.

NOTES

1. Cornel West, *Prophesy Deliverance! An Afro-American Revolutionary Christianity* (Philadelphia: Westminster Press, 1982), 84–85; Clarence E. Hardy III, *James Baldwin's God: Sex, Hope, and Crisis in Black Holiness Culture* (Knoxville: The University of Tennessee Press, 2003), 105.
2. This chapter is being published by Punctum Press in *Meaningful Flesh: Reflections on Religion and Nature for a Queer Planet*, ed. Whitney Bauman, 2017.
3. Michael, Warner, ed. *Fear of a Queer Planet: Queer Politics and Social Theory* (Minneapolis: University of Minnesota Press, 1993), xxvii.
4. Carol Wayne White, *Black Lives and Sacred Humanity: Toward an African American Religious Naturalism* (New York: Fordham University Press), 95.
5. Cora Kaplan and Bill Schwartz, *James Baldwin: America and Beyond* (Ann Arbor: The University of Michigan Press, 2011), 3.
6. James Baldwin, *Baldwin: Collected Essays,* ed. Toni Morrison (New York: The Library of America, 1998), 810.
7. Frederick Douglass, *Selected Speeches and Writings,* ed. Philip S. Foner and Yuval Taylor (Chicago: Chicago Review Press, The Library of Black America Series, 2000), 269.
8. Baldwin, *Collected Essays*, 137.
9. Ibid., 119.
10. Ibid., 124.
11. Ibid., 7–8.
12. James Baldwin, introduction to *The Price of the Ticket: Collected Non-Fiction, 1948–1985* (New York: St. Martin's Press, 1985), xx.
13. Baldwin, *Collected Essays*, 468.
14. Ibid., 8.
15. Ibid., 810.
16. Ibid., 125.
17. James Baldwin, “No Name in the Street,” in *The Price of the Ticket*, 410–11.
18. Ibid., 380.
19. Baldwin, *Collected Essays*, 377.
20. Baldwin, *The Price of the Ticket*, xix.

21. James Baldwin, *The Cross of Redemption: Uncollected Writings* (New York: Vintage, 2010), 200.
22. Fred Standley and Louis H. Pratt, *Conversations with James Baldwin* (Jackson: University Press of Mississippi, 1989), 50.
23. Baldwin, *Collected Essays*, 302.
24. Baldwin, *Cross of Redemption*, 196.
25. Baldwin, *Collected Essays*, 305–5.
26. Ibid., 319.
27. Ibid., 309.
28. Ibid., 341.
29. Ibid., 366.
30. Ibid.
31. Standley and Pratt, *Conversations*, 48.
32. Baldwin, *Cross of Redemption*, 203.
33. James Baldwin, "Notes for a Hypothetical Novel" in *Nobody Knows My Name* (New York: Vintage), 141–52.
34. D. Quentin Miller, ed. *Re-viewing James Baldwin: Things Not Seen* (Philadelphia: Temple University Press, 2000), 3.
35. Baldwin, *Collected Essays*, 334.
36. Peter Van Ness, *Spirituality and the Secular Quest* (New York: Crossroad, 1996), 5.
37. Donald Crosby, *Living with Ambiguity: Religious Naturalism and the Menace of Evil* (Albany: State University of New York Press, 2008), ix–x.
38. Loyal Rue, *Religion Is Not about God: How Spiritual Traditions Nurture Our Biological Nature and What to Expect When They Fail* (Piscataway, N.J.: Rutgers University Press, 2005), 77.
39. Ibid., 75.
40. White, *Black Lives*, 33ff.
41. Ursula Goodenough, *The Sacred Depths of Nature* (New York: Oxford University Press, 1998), 75.
42. Standley and Pratt, *Conversations*, 54.
43. Ibid., 55.
44. Baldwin, *Collected Essays*, 816.
45. Ibid., 824.
46. Ibid., 828.
47. Ibid.
48. Claudia Schippert, "Implications of Queer Theory for the Study of Religion and Gender: Entering the Third Decade," *Religion and Gender* Vol. 1, No. 1 (2011): 66–84.
49. Ibid., 74.
50. White, *Black Lives*, 125–6.
51. Ibid., 76.
52. Baldwin, *Collected Essays*, 841.

Chapter Six

Slenderman

A Trans Hermeneutic of the Apocalypse

Teresa Hornsby

Most people live in a fantasy. They deny that the world is chaotic; they think that world fits into neat categories; they believe that there is an "order of things." And these people assume that this supposed order, these categories, are established at a biblical-like creation, and thus, they are "natural." And at the foundation of all order, of all neatness and compartmentalization, is the gender binary. However, practically all of us have deluded ourselves into accepting that gender is something that is static and fits neatly into an either/or of male/female. This gender binary is what holds the rest of the world together by creating the hierarchies that seek to provide stability and security. Capitalism can only exist if some bodies are worth less than other bodies. For example, Western religious institutions prevail only if there is imperfection longing for perfection; medical and mental institutions need the sick, prisons need prisoners, and so on. All of these "lesser" elements of any binary relationship are feminized.[1] But . . . the world isn't neat and tidy. Things do not fit into two defined units; things get messy, ambiguous; there is blood, decay; we suffer, and we die.

Historically, we of Eurocentric heritage target, hate, and fear those things or persons that embody ambiguity. It's how and why we create our "monsters."[2] One manifestation of this fear is the extraordinarily high levels of violence enacted upon bodies who do not "fit" into neat categories: interracial individuals, immigrants, and particularly trans bodies; they are expressions of a collective terror of ambiguity. Yet, there are new constructions of horror, horror that is produced in a new and present apocalyptic crisis, and these horrors show signs of hope: the font from which the terror flows is no longer "ambiguity" but a more tangible (or "real") threat. The "open-source" produced "Slenderman" is such a monster.

Slenderman lurks. Freakishly tall, clad in a black and white business suit, white, bald, and faceless, he preys on the innocent. Slenderman is the monster of this new, as yet unnamed, generation. He inherits the facelessness of Jason (the *Halloween* series) and the ethereal nature of Freddie (*Nightmare on Elm Street)*, but his birth and construction are a phenomenon in the art of narratival monster building called "Open Sourcing." Slenderman started simply in 2009 as a few photographs submitted to an online forum called "Something Awful;" the group invited users to submit an image of pure horror. A participant who called himself "Victor Surge" uploaded photoshopped pictures of children playing in a schoolyard with a mysterious figure lurking in the shadows behind; the next photo showed them running toward the camera, their faces betraying sheer terror. In her article "Open-Sourcing Horror," Shira Chess writes:

> On 10 June, the tenor of the forum shifted dramatically, though, when a user posted two "photos" and a news story identifying a faceless "Slenderman" in a suit who stalked children. Almost immediately, an obsessive interest in the Slenderman took over the forum discussions. Near constant additions were added to the Slenderman mythos with new photographs, drawings, short fiction, and even wood cuttings showing his appearance in different times and places.[3]

From these fragmental beginnings, Slenderman has become a paranormal terror that exists in the shadows, stalks small children, often appearing in their dreams before they disappear. His origins are ambiguous; his identity shifts; he is faceless and nameless. His construction has been the result of thousands of online users creating fake documentary footage, photos, games, news items, blog entries, and more. The primary locus of this construction is a YouTube series called, "Marble Hornets," the story of a film student named Alex who is shooting a film in the summer of 2009. As the series progresses, Alex becomes paranoid, irritable, and abruptly stops filming. He claims that the working conditions have become impossible and threatens to burn all of the videos. Alex disappears and a friend decides to go through each film to determine the cause of Alex's deteriorating mental state. Each of the videos is one episode of Marble Hornets; there are 135 total.

Though its origins are humble, the first episode "Introduction" and the first Slenderman entry on the "Marble Hornets" YouTube channel have over 12.3 million views.[4] Other videos (a fake documentary, a short film, pseudo news reports, police recordings, etc.) have millions of views for each site.[5] The open-sourced construction from which The Slenderman ascended is a phenomenon of itself: thousands take part in the creation, millions watch. The mythos of Slenderman began with eerie fake "eye witness" photographic accounts, such as this entry by Surge:

> "we didn't want to go, we didn't want to kill them, but its persistent silence and outstretched arms horrified and comforted us at the same time . . ." 1983, Photographer Unknown, assumed dead.[6]

Horrific indeed. The Marble Hornets videos are chilling, eerily unsettling, but at the same time, enticing. What is it that draws *millions* to such a simple tale? Being "horrified and comforted" is exactly the reaction that Julia Kristeva observes in each encounter with the abject—the point at which "the real" dissolves the fictions that we have created to contain it. The abject is that which cannot be subsumed into order; it cannot be categorized, or managed in service to structure and security. The abject erupts at the borders—the areas that cannot be defined, contained, or restrained. It is:

> death infecting life. Abject. It is something rejected from which one does not part, from which one does not protect oneself as from an object. Imaginary uncanniness and real threat, it beckons to us and ends up engulfing us . . . It is, what disturbs identity, system, order. What does not respect borders, positions, rules.[7]

> It lies there, quite close, but it cannot be assimilated. It beseeches, worries, and fascinates desire, which, nevertheless, does not let itself be seduced. Apprehensive, desire turns aside; sickened, it rejects.[8]

The tales of The Slenderman, I think, represent the anxieties of the abject as they emerge through the genre of open-sourced horror; this is the type of fiction that can do justice to the terror that erupts when the abject threatens the flimsily produced edifice of safety and security: the source is ethereal; there is no static substance; and its signification is open and dynamic. The Slenderman and its productive apparatus stand in the place of a gaping void—a *tĕhôm* onto which we place our present anxieties. This *tĕhôm,* the void that we encounter in Genesis, and upon which we are perched precariously, simultaneously drawn in and repelled by the irresistible vertigo of staring into the chasm, expresses the abject, the chaotic uncertainty that comes with the growing realization that there is no ground of being, that there is only instability, and that this instability and uncertainty have always been a part of creation. Yet, the distinction is that these anxieties, laid bare in creation, are now projected into the future.

These anxieties arise, as Mary Douglas brilliantly illustrates, because Western cultures seek to delineate differences by dualistic categorization—everything must fit neatly and precisely into one and only one group; these categories were laid out in the creation of the world. Any ambiguity is perceived as unnatural, monstrous, a threat, and, of course, dangerous.[9]

Douglas illustrates her ideas through the example of the Levitical codes in the Hebrew Bible. She argues that the Israelites understood themselves to be one body, and by governing the orifices (mouth and genitals) and the borders (the skin) of that body, they were, in fact, shoring up any potential points of community vulnerability. The implication being, that any infraction of laws that govern the body (which is a symbolic microcosm of the Israelite world) opens up the entire community to potential destruction. The prophets then, understood that the invasion, destruction, and forced exile of Israel (and Judah) to be consequences of neglected and eroded boundaries. The apostle Paul understands the body's boundaries and the fledgling Christian community's borders in exactly the same way.[10]

What is then assumed about the fixedness of gender is that gender is determined by nature (or God) and is inextricably connected to the body (which is also fixed). Thus, it follows, that if we believe that the two genders are "natural," or "God Given," then any expression of gender that does not conform to the predetermined sets of actions allowed to that body, those expressions are considered to be odd—but not a neutral or benign odd (if there is such a thing), but one that is morally weighted.[11] As Judith Butler famously writes, "indeed, those who fail to do their gender right are regularly punished."[12] "Performing right," usually means, first and foremost, having sexual desire for the opposite sex. Everything else stems from that desire, for example, wearing the correct clothing, proper adornment of one's hair, eating the correct food, riding the appropriate bicycle, or viewing television or films suitable to one's gender. All others are "monstrous."

The cultural processes of open-sourcing, especially as it intersects with the genre of horror, allow us to examine, clearly and concisely, how and why monsters are made. Open sourcing erupted in the 1990s, following the free software movement of the 80s.[13] The distinctions of creating fiction in this way are that the processes are open to anonymous collaboration; it is dynamic, which means that while it maintains certain core pieces (for example, the freakish height and dark suit of Slenderman), the constructions are fluid. It remains without definition, as it is modifiable in infinite ways.

While open sourcing may not seem that distinct from the constructions of fictions in general, the fact that these productions are done online, and with unfamiliar collaborators with little to no hard-fast rules in place, provides insight into the ways in which the creation of monsters parallels cultural anxieties and attends to social needs.

Horror, according to Kristeva, is the privileged signifier of the thing that terrifies us most, the abject. The abject appears when the realities of our own mortality (decay, disease, death) seep through the order that we have constructed for ourselves. Those anxieties and fears that erupt can only be expressed through horror, the "ultimate coding of our crises" (1982, 208). Confrontations with the abject terrorize us because they lay bare the ambiguities, thus the instabilities, upon which we place everything. Horror, as Kristeva explains, is when we are confronted with an ambiguity that will not allow us to deny the reality of death. "Abjection is above all ambiguity. Because, while releasing a hold, it does not radically cut off the subject from what threatens it—on the contrary, *abjection acknowledges it to be in perpetual danger*."[14]

In the Christian canon, these horrors tend to be expressed through the genre of the apocalypse. The mythos of The Slenderman is not of the genre of apocalyptic literature *per se*,[15] but is produced in an atmosphere that exhibits the same anxieties as those that have historically produced an apocalypse. The apocalypse typically addresses some imminent crisis, and "evokes the reign of chaos, which renders meaningless concepts such as justice and goodness."[16] An apocalypse is an intertwined matrix of two purposes: to describe the social situation at hand and to suggest a strategy that would relieve the stresses of present evil.[17] The genre

> revealed not only the future course of history, or the final disposition of mortals, they also unveiled—that is, interpreted—present reality. Monsters characterize imperial brutality; cosmic portents reflect social injustice; heavenly glories display the rule of the transcendent over the ordinary.[18]

In other words, our horrors reenact the confusion that we experience when nothing makes sense: What is "right"? To what are we tethered? Who or what demands and enforces Justice? Chaos. No stability. No security.

Here is the essence of our present, postmodern horror: our "realities" are contingent upon a dualist paradigm, and that paradigm depends upon a two, and only two, gender system; it is upon that dualism that all order stands. To break the imaginary two-gender paradigm is to destroy the foundation of all order. In other words—all comfort, security, order, logic, sense—all rest upon . . . nothing.

Thus, the "monsters" are those things that we imagine threaten that stability—they are the entities that do not fit neatly into a binary—the zombie, the vampire, Frankenstein's monster (neither alive nor dead), the werewolf (neither human or non-human animal). We see in the BBC America produced *Orphan Black,* a series about corporate-produced clones with distinct personalities, that even in the "hard" sciences of evolutionary developmental biology (evo-devo), genetic modification, genome sequencing, and, subsequently, cloning (Frankenstein 2.0?), a goal is to eliminate flaws, to create a more perfect subject, and therefore to introduce more order and stability into a flawed creation. Yet, as is often the case in *Orphan Black,* the product tends to be monstrous. Even in our imaginary utopias, chaos and all of the matters that cannot be contained in tidy compartments continue to seep. The monstrous is present in the beginning, in the duration, and in the end. The hard sciences would do well to acknowledge and remember this fact.

And in our current state of hyper binarism, the trans person's body has become ambiguity personified. The trans person, as Julia Serano names, "a whipping girl," has become a target for all woes, and is often portrayed as monstrous, and thus, in our present apocalyptic crisis, is perceived to be deserving of the extraordinarily high rates of violence. I think the heightened rate of violence is due to the fact that a trans person is a flesh-and-blood, incontrovertible proof that a stable and naturally created gender binary is imaginary; the trans body is the ambiguity that remains connected to the subject, that is, the abject that cannot be subsumed into the imaginary. However, the abject, or the "monstrous," is not (as Kristeva tends to imply) something that is apart from a carefully constructed worldview, but should be viewed, as Judith Butler and Margrit Shildrick have argued, as part of the constructed order. Abjection, chaos, the monstrous, ambiguity—whatever the term—is an essential element of a joyous existence and is present even in biblical creation.[19] Yet, the hierarchical, imaginary dualism upon which Western civilization stands renders those realities (abjections) as demonic, horrific, and dangerous. If we could read Biblical texts without the assumptions brought on by a gender binary, we would read the canonical apocalypse with all of its abjection, fluidity, seepage, and transitions intact. The fact that the Christian canon concludes with what could arguably be the most liminally abjective, queer writing in the entire Bible confirms, for me, that we are heading in the right direction when we recognize the dynamism of the text, and its refusal to be subsumed into hierarchical dualism. Tina Pippin is one scholar who has already recognized the possibilities of a trans hermeneutical reading of Revelation.

Pippin remarks that in John's Apocalypse, "nothing is stable . . . , especially gender and desire."[20] Pippin observes that 144,000 males become the Bride of Christ, and "by becoming women, men can love a man (God) without the threat of homosexuality."[21] Thus, the canonical apocalypse cannot function without transvestitism and homoeroticism as its central tropes.

Actually, one should expect the canon to conclude in this way, especially if we take into account that Jesus is a site where so many dualisms are dissolved. Jesus' ambiguity exists usually not as a result of any one portrayal of him, but as a result of how the reader imagines Jesus after combining all the evangelists' portraits. Mark's human Jesus exists simultaneously with John's ethereal Logos. Theologian Robert E. Goss reflects upon the "Queer Fluidity" in his commentary on John's Gospel:

> John's Gospel develops the most profound Christological reflections upon Jesus that we encounter in the Christian canon of scriptures . . . It weaves the gender

> fluidity of Jesus as the embodiment of Divine Wisdom with Jesus as God's revelation. Jesus is the supreme manifestation of God's Word and embodiment of Divine Wisdom. It expresses an alternative to the male imagery of God.[22]

Goss goes on to quote Martin Scott, saying,

> The point of John's Wisdom Christology is precisely that Jesus Sophia is not mere man, but rather the incarnation of both the male and the female expression of the divine, albeit within the limitation of human flesh.[23]

In addition to a fluidity of gender, Jesus' body becomes the site where multiple dualisms collapse: divinity and humanity merge; "this world" blends into "other world"; and through the resurrection, physicality and ethereality are one. It isn't an exaggeration to say that all things "trans" lie at the heart of Christian belief: the *trans*formation of the resurrected corpse; the *trans*ubstantiation of bread to body, of wine to blood; and of course, the *trans*ition of God to human to God. John's apocalypse, then, holds a representation of Jesus as a site of what Laura Jane Grace (a trans front woman for the punk group "Against Me!") calls "the true trans." Jesus is a "true trans" in that as he appears in this apocalypse, dichotomies are collapsed: physical/spiritual, human/divine, male/female, I/other, human/animal, alive/dead, etc. He exists in the interstices of any imagined order; he becomes a personification of the abject body. Thus, a trans hermeneutic of the apocalypse recognizes that one cannot do justice to the final book of the canon unless the apocalyptic Jesus and his court of transvestite brides are accepted as the queer and chaotic beings they are. However, I must be clear here: the Christian apocalypse is basically a trans-phobic expression as it conveys a debilitating fear of ambiguity and fluidity. Ironically, as Pippin notes, its mode for articulating those fears is through trans bodies and ambiguous sexualities, yet the phobias are never resolved and in the violent end, the dualistic and hierarchical constructs are left in place.

Likewise, the apocalyptically and virtually produced tale of The Slenderman betrays our collective fear of chaos and anxieties that originate with ambiguity and the failures of a system that hopelessly tries to stabilize them. What is striking about The Slenderman is the site upon which all fear is placed: a tall, white, faceless male in a business suit who comes after our children; to quote Pippin, "the evil powers are in plain view now."[24]

As my students and I discussed The Slenderman, we pondered the question, what is scary about a white, aging man in a business suit? Our conversations turned to horrifically "real" evils brought about by "men in business suits." We talked about Wall Street, rampant corporate greed that has led to human trafficking, the destruction of natural resources, the colonization of indigenous people . . . in other words, a neoliberal capitalist ideology which is sustained by a gender binary that creates expendable bodies. A gender binary renders some bodies less valuable than other bodies, which creates cheap labor, which produces profit.[25]

The promise that modernity, neo-liberal capitalism, and Fordism (an industrial-based form of capitalism) would guarantee a safe and secure world is empty. The symbol of twenty-first century, colonial power (white, cis-male, capitalist) has become the monster that haunts our waking life and threatens the lives of our children. Simply put: The Slenderman is a personification of a neo-liberal, capitalist evil.

The second century horror, John's Revelation, expresses—violently, grotesquely—the very real terror of that time and that place: a dissolution of boundaries and their subsumption into an evil system.[26] These fears are personified in the image of demonic and feminized sexuality, the whore of Babylon. Their resolution is found, partly, in ambiguous sexuality and trans bodies, that is, a seepage of the abject into the fantasy, yet this horror ends in genocide. The twenty-first century horror, The Slenderman, betrays the same anxieties (fear of ambiguity, a fear of the abject, and the dissolution of the subject), yet there is something new here. The site from which all terror originates in this tale is, indeed, a very "real" monster. It is a monster that exists among us, has sponsored genocide, the destruction of whole civilizations, and the enslavement of millions. And though this horror is still in process, there is hope here. Though the anxieties may have emerged from abjective concerns (it is the programmed ideology after all), the significance of the fact that this horror is transitive and multi-vocatively produced cannot be overstated. Open-sourced horror, in this case, becomes an expression of the abject; it is the means by which the abject is given representation and subjectivity. More, the "monster," The Slenderman, reveals the logical end, the "real" consequences of accepting an imaginary and dualistic order, and NOT embracing ambiguity, discursiveness, and bodies in transition. The canonical apocalypse is a failed and hopeless vision that ends in death. But this new apocalyptic response, one that embraces ambiguity, a trans hermeneutic if you will, knows the "real" threat that brings our end—a neo-liberal capitalism born of greed and ignorance, and this hermeneutic imparts a means of survival: to embrace the ambiguous and purge the illusion of a gender binary. No more binary, no more monsters.

WORKS CITED

Althaus-Reid, Marcella. *Indecent Theology Theological Perversions in Sex, Gender and Politics*. London: Routledge, 2001.

———. *The Queer God*. London: Routledge, 2003.

Brooten, Bernadette. *Love Between Women: Early Christian Responses to Female, Homoeroticism*. Chicago: Chicago University Press, 1996.

Butler, Judith. "Performative Acts and Gender Constitution: An Essay in Phenomenology and Feminist Theory." *Theatre Journal* 40 (1988): 519–53.

———. *Gender Trouble. Feminism and the Subversion of Identity*. London: Routledge, 1990.

———. *Bodies That Matter: On the Discursive Limits of 'Sex.'* New York: Routledge, 1993.

———. "Against Proper Objects." *Differences: A Journal of Feminist Cultural Studies* 6 (1994): 1–26.

———. *Undoing Gender*. New York: Routledge, 2004.

———. "Doing Justice to Someone." In *Transgender Studies Reader*. Edited by Susan Stryker, and Stephen Whittle, 189–93. New York: Routledge, 2006.

Carey, Greg. *Ultimate Things: An Introduction to Jewish and Christian Apocalyptic Literature*. St. Louis, MO: Chalice Press, 2005.

Chesler, Phyllis. *Women and Madness*. New York: Doubleday, 1972.

Chess, Shira. "Open-Sourcing Horror." *Information, Communication, and Society* 15 (2012): 374–393.

Collins, John J. *The Apocalyptic Imagination: An Introduction to the Jewish Matrix of Christianity*. New York: Crossroad Publishing, 1987.

Countryman., William. *Dirt, Greed and Sex: Sexual Ethics in the New Testament*. Minneapolis: Fortress Press, 1988.

Douglas, Mary. *Purity and Danger*. London: Routledge, 1966.

Foucault, Michel. *A History of Sexuality. Vol. 1: An Introduction*. New York: Pantheon, 1978.

Goss, Robert E. "John," In *The Queer Bible Commentary*, edited by Deryn Guest, Robert E. Goss, Mona West, and Thomas Bohache, 548–65. London: SCM Press, 2006.

Guest, Deryn. "Troubling the Waters: *Tĕhôm,* Transgender, and Reading Genesis Backwards." In *Transgender, Intersex, and Biblical Interpretation*, 21–44. Edited by Teresa J. Hornsby and Deryn Guest. Atlanta: Scholars Press, 2016.

Hennessy, Rosemary. *Profit and Pleasure: Sexual Identities in Late Capitalism*. New York: Routledge, 2000.

Hornsby, Teresa J and Guest, Deryn. *Trans Gender, Intersex, and Biblical Interpretation*. Atlanta: Scholars Press, 2016.

———. "Capitalism, Masochism, and Biblical Interpretation." In *Bible Trouble: Queer Reading at the Boundaries of Biblical Scholarship*, 301–320. Edited by Teresa J. Hornsby and Ken Stone, Atlanta: Society of Biblical Literature, 2011.

Huber, Lynn. "Gazing at the Whore: Reading Revelation Queerly." In *Bible Trouble: Queer Reading at the Boundaries of Biblical Scholarship*, 301–320. Edited by Teresa J. Hornsby and Ken Stone. Atlanta: Society of Biblical Literature, 2011.

Kristeva, Julia. *Powers of Horror: An Essay on Abjection*. European Perspectives. Translated by Leon S. Roudiez. New York: Columbia University Press, 1982.

Lorber, Judith and Lisa Jean Moore. *Gender and the Social Construction of Illness.* Plymouth, UK: AltaMira Press, 2002.

Martin, Dale B. *The Corinthian Body*. Reprint. New Haven: Yale University Press, 1999.

——. *Sex and the Single Savior: Gender and Sexuality in Biblical Interpretation.* Louisville: Westminster John Knox, 2006.

Novak, Michael Anthony. "The Odes of Solomon as Apocalyptic Literature." *Vigiliae Christianae* 66 (2012): 527–550.

Pippin, Tina. "The Joy of Apocalyptic. Sex." In *Gender and Apocalyptic Desire*, 64–75. Edited by Brenda E. Brasher and Lee Quinby. London: Equinox, 2006.

Pippin, Tina and J. Michael Clark. "Revelation/Apocalypse," in *The Queer Bible Commentary*, 753–68. Edited by Deryn Guest, Robert E. Goss, Mona West, and Thomas Bohache. London: SCM Press, 2006.

Pippin, Tina. *Apocalyptic Bodies: The Biblical End of the World in Text and Image.* London and New York: Routledge, 1999.

Shildrick, Margrit. *Embodying the Monster: Encounters with the Vulnerable Self.* London: Sage, 2002.

Online Sources for Slenderman:

Gearmark.tv. "Fathom: Slenderman." YouTube video. Posted on Oct. 20, 2013. https://www.youtube.com/watch?v=q2duevcUMGw

fxscreamer. "Slender Suits: The Documentary of Slenderman." YouTube video. Posted on Oct. 28, 2010. https://www.youtube.com/watch?v=ysgQ9spsphk.

Marble Hornets: all episodes: https://www.youtube.com/watch?v=Wmhfn3mgWUI&list=PLCACDEE1E71FC8E5E

NOTES

1. Following Foucault's thesis (in Michel Foucault, *A History of Sexuality. Vol. 1: An Introduction* (New York: Pantheon, 1978)) that power is produced and maintained via social institutions, subsequent work makes it clear that integral to each institution's existence is a hierarchized gender binary. For economic theory see Rosemary Hennessy, *Profit and Pleasure: Sexual Identities in Late Capitalism* (New York: Routledge, 2000); for the gendering of religious ideologies see Marcella Althaus-Reid, *Indecent Theology Theological Perversions in Sex, Gender, and Politics* (London: Routledge, 2001); for the feminization of illnesses see Judith Lorber and Lisa Jean Moore, *Gender and the Social Construction of Illness* (Plymouth, UK: AltaMira Press, 2002); and Phyllis Chesler, *Women and Madness* (New York: Doubleday, 1972).

2. See Guest's discussion of Shildrick's *Embodying the Monster: Encounters with the Vulnerable Self,* in *Trans Gender, Intersex, and Biblical Interpretation,* eds. Teresa J. Hornsby and Deryn Guest (Atlanta: Scholars Press, 2016), 21–44.

3. Shira Chess, "Open-Sourcing Horror," *Information, Communication, and Society* 15 (2012): 375.

4. MarbleHornets. "Entry #1." *YouTube.com.* Last modified. January 20, 2009. https://www.youtube.com/watch?v=Bn59FJ4HrmU. As of 12/20/2016 this short film has 5.9 million views.

5. Gearmark.tv "Fathom: Slenderman," YouTube video, posted Oct. 20, 2013, https://www.youtube.com/watch?v=q2duevcUMGw (as of 12/20/2016, this short film has 9.9 million views); fxscreamer. "Slender Suits: The Documentary of Slenderman," YouTube video, posted Oct. 28, 2010. https://www.youtube.com/watch?v=ysgQ9spsphk (as of 12/20/2016, this short film presently has just over 8 million views). There are 135 videos that make up the Slenderman Marble Hornets network.

6. Chess, 378.

7. Julia Kristeva, *Powers of Horror: An Essay on Abjection*, European Perspectives, trans. Leon S. Roudiez (New York: Columbia University Press, 1982), 4.

8. Ibid., 1.

9. Mary Douglas, *Purity and Danger* (New York: Routledge, 1966).

10. As it has developed out of Levi-Strauss' structuralist anthropological theory, Mary Douglas' scholarship on boundary construction provides a foundation for recent work on the intersections of gender and New Testament. Two studies that expound on Paul's community building and boundary concerns are: William Countryman, *Dirt, Greed and Sex: Sexual Ethics in the New Testament* (Minneapolis: Fortress Press, 1988); Dale Martin, *The Corinthian Body* (New Haven: Yale University Press, 1999).

11. Bernadette Brooten, *Love Between Women: Early Christian Responses to Female, Homoeroticism* (Chicago: Chicago University Press, 1996) provides a thorough exploration of first century attitudes regarding gender and sexuality.

12. Judith Butler, "Performative Acts and Gender Constitution: An Essay in Phenomenology and Feminist Theory," *Theatre Journal* 40 (1988): 521.

13. Chess, 381.

14. Kristeva, 9.

15. The characteristics of the apocalypse are as follows: it is grim and dark (Michael Anthony Novak, "The Odes of Solomon as Apocalyptic Literature," *Vigiliae Christianae* 66 [2012] : 527–550); it usually portrays a venerable human guided by a supernatural force that is depended upon for interpretation (John J. Collins, *The Apocalyptic Imagination: An Introduction to the Jewish Matrix of Christianity* [New York: Crossroad Publishing, 1987], 5–6); it is an "elaborate review of history, presented in the form of a prophecy and culminating in a time of crisis and eschatological upheaval" (Collins, 6). It is a final judgment, with a destruction of the wicked and a hortatory aspect (Collins 1987, 6). Collins writes, "The world is mysterious and revelation must be transmitted from a supernatural source, through the mediation of angels; there is a hidden world of angels and demons that is directly relevant to human destiny; and this destiny is finally determined by a definitive eschatological judgment. In short, human life is bounded in the present by the supernatural world of angels and demons and in the future by the inevitability of a final judgment" (8). Greg Carey lists the following four characteristics (following Collins): "the expectation of a final judgment, in which God separates sinners from the righteous; hope for resurrection of the righteous dead to a glorious realm; reflection upon God's role in history, both

past and future, leading to a new age of justice and deliverance; and speculation concerning a heavenly messianic figure—a Son of Man, Elect One, Righteous One, or Messiah—who will administer final justice upon the world" (Collins, 19).

16. Greg Carey, *Ultimate Things: An Introduction to Jewish and Christian Apocalyptic Literature* (St. Louis, MO: Chalice Press, 2005), 228.

17. Collins, 32.

18. Carey, 228.

19. See Deryn Guest, "Troubling the Waters: *Tĕhôm,* Transgender, and Reading Genesis Backwards," in *Transgender, Intersex, and Biblical Interpretation*, eds. Teresa J. Hornsby and Deryn Guest (Atlanta: Scholars Press, 2016), 21–44.

20. Tina Pippin, "The Joy of Apocalyptic. Sex," in *Gender and Apocalyptic Desire*, edited by Brenda E. Brasher, and Lee Quinby, 64–75 (London: Equinox, 2006), 72.

21. Ibid., 71.

22. Robert E. Goss, "John," in *The Queer Bible Commentary*, edited by Deryn Guest, Robert E. Goss, Mona West, and Thomas Bohache (London: SCM Press, 2006), 550.

23. Goss, 550.

24. Pippin, *Apocalyptic Bodies: The Biblical End of the World in Text and Image* (London and New York: Routledge, 1999), 79.

25. Hornsby, "Capitalism, Masochism, and Biblical Interpretation," in *Bible Trouble: Queer Reading at the Boundaries of Biblical Scholarship*, edited by Teresa J. Hornsby and Ken Stone (Atlanta: Society of Biblical Literature, 2011), 301–320.

26. Pippin, "The Joy of Apocalyptic. Sex," 72.

Chapter Seven

"Nothing in This World Is Indifferent to Us"

A Dialogical Reflection on the Queerness of Theology and Science

Catherine Keller and Laurel Schneider

This essay is a collaboration based on other collaborations. Catherine Keller and Laurel Schneider have thought, spoken, and written together in various ways over the years. This dialogue essay is the result of a process that began with a line that Laurel pulled from a document that Catherine suggested they read together: "Nothing in this world is indifferent to us." Musing on this line in terms of its relevance for queer theory's recent turn to affect, feminist theology's historic concern with divine relationality, and scientific investigations of quantum uncertainty led to talking, then writing, then passing of passages back and forth, then meeting and continuing to talk in the context of the conference that inspired this book. This is therefore a dialogue that explicitly embodies a process of exchange and intellectual growth. It continues of course, rhizome-like, in the present. No surprise, really. Process thinking is a hallmark of much of our individual and shared theological and philosophical work. And it informs much of our interest as theologians in what is happening currently at the cutting edges of scientific inquiry. Queer theory, theology, and science have much to offer each other and appear, already, to be in deep relation.

A QUEER MATTER OF THEOLOGY AND SCIENCE

* * *

LCS: "Nothing in this world is indifferent to us." These words capture, on one level, the theologies we pursue, namely theologies with planetary sensibilities. These words also indicate the sciences that capture us, as theologians. We have encouraged ourselves to listen more carefully to those sciences for some time now, to attend more closely to scientific explorations as the discoveries widen and the mysteries therein deepen. We do this because we fundamentally agree that nothing in this world is indifferent to us. An indifferent world—a world where differences literally and philosophically don't matter is a world incapable of incarnation. Incarnation after all is the idea central to Christianity that divinity is enfleshed in the actual material of the world, which by necessity means the actual differences that constitute the world. For many Christians, this enfleshment of divinity is particularly revealed in the differences that one specific body in first-century Palestine made and makes. This means that Christian theology that takes incarnation *seriously* cannot consider the cosmos and everything we continue to learn of its complex interconnections indifferently. Of course, there are many so-called Christian theologies that avoid the actual messiness and materiality of bodies even in the midst of extolling the doctrine of incarnation. They do this by reserving the category of incarnation to one human body, that of Jesus of Nazareth. Such theologies imagine that the story of divine incarnation can be kept pure of everything that bodies entail—like sex, animality, hunger, waste, quantum entanglements, DNA and its inter-species ambiguities, the Heisenberg uncertainty principle, and, in short, everything else that complicates bodies in a state of inexorable interconnection and multiplicity.

Theology that takes incarnation seriously in the sense of bodies themselves, that dwells in the richness of what "fully human, fully divine" must mean beyond the words of a doctrinal affirmation, is theology that is open to the ambiguity and complexity that bodies present. This makes incarnational theology particularly amenable to queer theory, as that theory also concerns itself with bodily interconnection, porosity, and complexity on the one hand and with productions of ideology, power, and identities on the other.

I have written that queer theory is a means, not an end, wherever it is applied.[1] While queer theory is first about sexuality, modern notions of sexuality also implicate and help to construct modern constructions of race, and vice versa. Indeed, how else have the "races" been configured and managed? The modern sex/race co-constitution is fertile ground for so much else in human imagining because it is the locus of the most historically prolific and imaginative confusion about bodies and identities.[2] Sex/race via queer theory is precisely the place we can learn about how narrowly—and predictably—we find what we are looking for when bodies are at stake, because we already believe we know what we are looking for. In other words, the ideological, cultural, and socially constructed bases of so-called "natural" gender, sex, and race conceptions illustrate with helpful clarity the ideological, cultural, and socially constructed bases of all human imagining about bodies—what early modern scientist Francis Bacon declared, back at the dawn of the modern scientific method, should be called the idols of the cave.[3]

As we have learned, queer theory helpfully interrogates the social career of binary genders, especially focusing on their participation in and production of racialized and nationalized structures of erotic power.[4] These are structures that sort and bestow lasting deleterious and salubrious effects on persons. Done right, queer theory helps unmask racialized and colonized genders; it gives a method for tracing the multiple ways that oppression follows the lines of flight of identity formation and categorization. It can also trace modern pathways through which identity-based oppression survives by a kind of viral morphology, feasting especially on identities forged in pain and rebellion, turning the marginalized into the disciplinarians of identity who are then capable of creating whole new categories of exclusion in the name of liberation.

Queer theory, perhaps only by the simple accident of starting out being *about* contradictions in gender, has opened up multiple ways of tracing this vast continuum of contradiction and co-construction in gender and race in the modern, colonial era (which is also the era of the birth of modern science). It helps us to understand that even the most banal of scriptures ("Abraham begat Isaac" say, or "for God so loved the world . . .") is implicated in interpretive complexities of gendered and eroticized meaning. This suggests that even the most banal of scientific equations, while ostensibly about simple phenomena, result in part from complexes of value that legitimate the experiments and calculations to begin with. Queer theory of this sort attends to seams and fissures in the *apparently* self-evident assertions we make. Not to make those assertions false, but to undermine their pretensions to finality. In the social sciences, queer theory helps us to see that often bricolage or assemblage is a better way for us to think about the inherent mobility of so-called natural identities or self-evident conclusions, especially their capacity to undergird domination dynamics.

But to do this work, like every tool of critical analysis queer theory needs material upon which to work, and that material is history and the complex social production of meaning and assumptions. Theology is therefore ripe for queer plucking because it is about identities and powers squared, namely the *divine* production of identities and powers. A simplistic and stereotypical rendering of theology's work (which unfortunately still exists in some quarters) is that theology is merely the explication of prior, and final, revelations handed "down" by a final and revealed (male) God. The fact that these hand-me-downs are themselves interpretations built upon millennia of cultural perception is, well, too often overlooked in those quarters. But we recognize that *all* theologies have always depended on the kindness of outside theories, scientific and otherwise. I submit that no intellectual enterprise, whether so-called scientific or so-called religious, is without such dependence on outside influences This is what Bacon surmised in the 17th century, it is what we know now. Odd that we all have such a hard time not stereotyping each other's passions and experiences, such a hard time remembering that we've always known that our endeavors are not pure of influence.

So what queer theory adds to contemporary work in physics and biology *and* theology is a theoretical attention to affects, norms, and the ways actual bodies attract, repulse, and otherwise exceed expectation and theory. For theologians, particularly those of us interested in divine-human relations, queer theory opens the category of relation to its own repressed and excluded realities. In other words, queer theory interrogates and finds new and living meaning in the very stuff theology holds in check.

In her book entitled *Indecent Theology*, Marcella Althaus-Reid helpfully focuses our attention on theology's participation in the production of status-quo norms of so-called decency.[5] What this usually means is the production of norms that exclude the so-called sinful, abominable, or simply indecent. Such exclusions are built upon more banal criteria of acceptability in any number of other everyday disciplinary mechanisms of conformity. Queer theory tells stories of indecency in order to illuminate the always wobbly, always implicated props of such practices of normalization. The Roman church's moves against the early scientists like Giordano Bruno or Galileo Galilei are hyperbolic examples of a more general and widespread social tendency toward disciplinary conformity. Edgy or "spooky" science (to use Einstein's term) also edges into queerness. Edgy theologies do the same, such as what James Cone did in his Black liberation theology when he pointed out the idolatrous whiteness in the church's cult of Jesus. These are acts of protest but also of attempted liberation. So queer theory reads the text (scripture, or hypothesis, doctrine or proof) again, with a particular taste for nonconformity that changes the subject of study forever and now and then opens up new possibilities for knowledge and practice. In theology queer theory sheds a different and discomforting sort of light on what we might mean by divine love and incarnation (particularly their persistent promiscuities). This does not mean that every discomfort or strangeness, or protest is queer. The discomforts are queer that reveal normative powers of exclusion. The strangenesses are queer that illuminate, in Althaus-Reid's terms, what dangles hidden but explicit beneath the robes of decency.

And so queer theory challenges every intellectual endeavor, particularly where it considers itself most decent. But I have also noticed the queer difficulty queer theory has with theology at the point of theology's own most dangerous and potentially indecent claims—of spirits, visitations, noumenal powers, and such. Queer theory has much to work with in the sciences and humanities so long as the question of spirits or of divine revelation never really arise. So long as "the religions" remain social artifacts among other social artifacts available to scientific exploration then queer theory can do its careful disentanglement of normalizing and disciplinary mechanisms at play in those artifacts. *Everything* in religion: the litanies, creeds and rituals, the historical narratives and declarations of faith, in a sense the *objects* of religion, can be queered by interrogation, their enmeshments in structures of power, their subversive survivals and fertile indecencies explored. But what has queer theory to say about theology's own faith/s in embodied spirits, its own ancient openness to post-Newtonian quantum relations and bodies (divine and otherwise) that engage promiscuously and surprisingly below, above, around—and in the closets of—any instrument that attempts to measure and observe what is "real"? Queer theory as an analytic method can scrutinize those claims for sure, connect them to social and political investments, and so on, but queer theory itself is challenged to its own edges of comfort when it comes to the possibility that what people experience as divinity might not be *merely* socially normed, imaginative projections but might include some actual—if always opaque—Other-in-relation. Queer theory is a useful check on hubris when theologians must *say* something about the biggest questions of planetary existence, of future hopes and past revelations, but the jury is still out on whether it can contribute much to those necessary claims. Queer theory is most comfortable poking fun of the ostentations of priestly (among which I include scientific) self-avowals of legitimacy. The jester is an important counter to the emperor, whatever discipline "His" pointy hat denotes. But queer *theology* can perhaps do more. At the very least it can revel in the possibilities of affective attractions or what Mel Chen calls animacies[6]— that throw everything from quarks to solar systems into a wonderful mash-up of relation and desire, including the possibility of a divine love that challenges from within every normative exclusion the religions have thought up.

Unless we cynically dismiss the various claims of theology (about God and the gods) as delusion, we are faced with an aspect of religion that establishes a limit for the kind of queer theory that only *un*masks ideological (and thus erotic) structures of power, and a limit for queer theories and for scientific theories that require totalized human (or other animal) agency at play in the structures of power and desire. In other words and somewhat paradoxically, religion understood queerly may limit queer theory—and every science of inquiry—much in the way that religions have always done to theory, scientific and otherwise.[7] But this time, theology in touch with incarnational thinking does so by being more queer than queer theory, and more open to scientific surprise at every level than the science/religion binary allows. Queer theory has to figure out how to say this—how it is—maybe—not yet queer enough for incarnational theology or for cutting edge ideas in physics.

So I close this part of our dialogue with Karen Barad's suggestion that there is a queerness already present in the structures of the world's matter by adding that there is queerness in "the religious" *by virtue of* its religiosity.[8] I think of Rudolf Otto's old notions of *mysterium fascinans* and *mysterium tremendum.*[9] These are two highly erotic notions in relation to each other—desire *and* repulsion, intimacy *and* spectral otherness, the inexorable pull of physicality to itself *and* awesomely abyssal separation. Both are mysteries of relation, applicable to the world we are beginning to see through the lens of the physicist, and through the camp glasses of the queer scholar, and through the old, old, eyes of the contemplative religious.

* * *

CK: *Tremendum et fascinans*: the mystery that stirs trembling—and at the same time attraction. Rudolf Otto's *Idea of the Holy* was one of the first works to study religion as a phenomenon, pursuing respectful enquiry into a multiplicity of religious truths and finding in all of them some version of this mystery that is awesome, awe-full, full of energy both attractive and withdrawing—quite an erotic push-pull . . .

That energy puts me in mind of a tension in queer theory itself. Consider José Esteban Muñoz's idea of "disidentification."[10] Writing as a queer of color, Muñoz on the one hand performs a radical withdrawal—*tremendum*—from any reified identity, hetero- or homo-normative. On the other hand, he offers in *Cruising Utopia* the complicating relationality of an irascible hope, the "not-yet" of a queer counter-public: *fascinans*! But does the *mysterium* then *already* queer religion? If religion in its deeper pulsation, beneath identification with this or that belief or community, evinces a certain queerness, then we might all here already be participating in it. Does that make us all queers? Well, no. That isn't what Laurel is saying; "queer" for us is not an identity in the way that lesbian or trans or straight might be. Nor would just anyone claim to be queer: I might say not that I *am* queer but that I am making use of a "queer methodology."[11] Or that I take part—*part,* partially, as participant—in the perspective of queerness.

We have been asked to consider together this disarming question: What is the relation between the possible queerness of religion and the possible queerness of science? Here is a quick answer: The relation between queerness and the religion-science relation comes down to *relationality itself.* But that says too much and therefore too little. More specifically, therefore, consider this: Religions lend expression to the ultimate mystery of things, and their more reflective representatives *know* mystery as something unknowable, invisible, unspeakable at the heart of the world. Therefore the contemplators and mystics of each religion are never content with the standardized talk of God—as all too knowable, indeed vocal and visible in the Abrahamic cases, in His sovereign masculinity.

Over the course of half a century, feminist theology has performed a great deconstruction of that gender, but we did not content ourselves with a new identification of the divine: as Goddess or She-God, as impersonal It, or as a relativized Him. We didn't, we don't, finally replace one fixed identity with another. We end up with a complicated metaphoric language, entangled in the burdens and the poetries of multiple traditions, of She/He/It/They pointing always beyond themselves to the unknown. And since gender is indissolvable from sex what is this sliding between and beyond genders if not, in a sense neither identitarian nor trivial, queer? Looking back we recognized in the tradition already all manner of intimacy of men with a male God, of same-sex intimacies between the religious, of poetic shifts of the language of God to that of the female beloved—both for men and for women mystics. The point is that in the forcefield of feminist theology we did learn to recognize God language—with either its normative masculinity or its sliding complexity—as demystifying means of relation to the mystery. Theology, both deconstructive and reconstructive, becomes explicitly and radically relational.

Such relationalism in feminist theology resonates all along with the complicating spectrum that then intensified gender as sex, that is diversifying and shifting and renamed sexes in the no doubt expanding alphabet of LGBTQ (and I and A and . . .) Queer theology as such is now still emerging in conjunction with queer theory, exceeding the denotations of gay and lesbian, and in alliances of activism. Clearly queerness is not just a new term for a transparent and bounded identity, as the noun "a queer" might suggest. Eve Sedgwick characterized its dynamism already in 1993: "A word so fraught as queer with so many social and personal histories of exclusion, violence, defiance, excitement . . . never can only denote; nor can it only connote; a part of its experimental force as a speech act is the way in which it dramatizes locutionary position itself."[12] Queerness is performing its positionality without vague-ing out into mere connotative metaphor—in which case anything strange can be sexed up as queer. It retains its crucial political reference to the specific histories of abjection and of eros—another kind of *tremendum et fascinans*.

So if this complicating queerness does bear some resemblance to the *mysterium tremendum et fascinans* of religious experience, or of "spirituality not religion," then the feminist and queer developments of theology can be said to intensify this spiritual push-pull, attraction-abjection. That is the "religion" part of the present conversation. But religion is always in danger of floating up, up, and away, into speculative theory if not into a supernaturalist theology—in spite of the vital matter of all these metaphors. That is one of the attractions of the religion-science conversation: it helps bring the spiritual energies of ancient and evolving traditions to bear upon material problems. But those of us not trained in the natural sciences also tremble at the risk of this discourse: not only must we fear being outed as inexpert or new age-y; there is also a substantive *tremens*, the fear that science, precisely through its high- and higher-tech force of *fascinans,* is emptying the world of mystery and meaning, abstracting from rather than relinking to a materiality that matters.

So then, speaking of matter, we find ourselves heading in the other direction: down, down, down, to the level of the little trembling, vibrating units of matter itself—to the quanta. It turns out that the quanta deconstruct any notion of a separate, fully known object just as surely—let me say just as knowably—as God gets deconstructed as the separate knowable subject. As Heisenberg put it: "The common division of the world into subject and object, inner world and outer world, body and soul is no longer adequate."[13] And of course once the dividing line between subject (me) and object (you)—observer and observed, in and out, soul and body—gets cloudy, it fogs up at a collective scale the established line between culture and nature. And therefore, of course, between religion and materiality.

In this challenge to the founding modern dualism—to what Whitehead called "arbitary disconnection of first principles"—we witness the slow 20th century undoing of the normative separation of reality into corporeal and mental substances. Natural normalcy collapses into "quantum weirdness" (a technical term among physicists, believe it or not). But quantum *queerness*? I admit to having attempted the phrase a few years ago, with a nervous tremble. But now I can quote with confidence a scientifically reliable source, Karen Barad, with her deep background in theoretical particle physics. I had discovered Barad's work however not because of its queerness, which does not explicitly characterize the physicist-philosopher's magnum opus, *Meeting the Universe Halfway*. Instead I found her while seeking support for my fascination with the mystery of quantum entanglement. Einstein had captured its *tremendum* in his phrase "spooky action at a distance." As the quantum physicist Shimon Malin put it, "It feels as if the ground is slipping under our feet, and we are not sure where we will land. This puzzlement and this feeling are characteristic of the first stage of a paradigm shift."[14]

And the shift entails, as Barad more lately demonstrates, some version of a relational ontology. She constitutes it primarily through her reading of Niels Bohr, for whom entanglements are the "components of phenomena." The phenomenon serves as the alternative to the commonsense presumption that we, alongside everything else, are separate subjects and objects that preexist our relationship to each other. Phenomena are not composed of elements that preexist each other; they come to be only together. Hence Barad prefers "intra-action" to "interaction" in designating this constituent relationality.

"For Bohr," she writes "phenomena—entanglements of objects and agencies of observation—constitute physical reality; phenomena (not independent objects) are the objective referent of measured properties."[15] In other words, the quantum intra-activity of observer and observed entangles our knowledge in the quantum process: "*we are part of that nature we seek to understand.*"[16] Barad recognized that "what is needed is a robust account of the materialization of all bodies 'human' and 'nonhuman,' and the contribution of all material forces, 'social' and 'natural.'"[17]

You can see why, for instance, Laurel Schneider, writing on "promiscuous incarnations" (don't miss its plural), might be theologically well supported by such an account.[18] Of course, for Barad the metaphor of entanglement, as the synchronistic interrelation of particles at any distance, retains its technical physical meaning. No matter how far it travels. "Phenomena" signify entanglements—not in the colloquial sense of a connection between individual entities separately preexisting, but in the technical sense of quantum entanglements "agentially intra-acting." Or, as she writes, "*Phenomena are material entanglements enfolded and threaded through the spacetimemattering of the universe.*"[19]

Hear how near to our enquiry Barad's metaphor travels: " . . . given that queer is a radical questioning of identity and binaries, including the nature/culture binary, this article aims to show that all sorts of seeming impossibilities are indeed possible, including the queerness of causality, matter, space, and time. Queer is not a fixed determinate term; it does not have a stable referential context, which is not to say that it means anything anyone wants it to be . . ."[20]

Queerness does not establish a new separable and knowable sexual identity but rather provokes the continual radical questioning of identity itself—and so also the deconstructions of those binaries that harden and enclose identity. Whether in traditional heteronormativities or in identity politics, whether in the macrosystems of the human or in the microsystems of the nonhuman, queer signifies indeterminacy all the way down. Down, that is, to the "queer quantum."

Here is performed what Emilie Townes calls "the destabilization of antagonistic dualisms."[21] However, just because the binaries go, it does not follow that just anything goes. As with quantum indeterminacy in Barad's reading of Bohr, the point is not that nothing is determined; rather, it is that there is an irreducible unknowability, indeed an unpredictability, built into the nature of things. Or, more precisely, into their material constitution, their ontology.

So we have to do not just with a matter of limits to our knowledge, behind which we may knowingly presume deterministic matter as usual along with all the normal causal relations between bodies we thereby expect. We might call that "uncertainty straight." Alternatively, "uncertainty *queer*" stakes the more radical claim: to an elemental open-endedness, spontaneity, and therefore unpredictability in all relations. The epistemic uncertainty is thus grounded in an ontological indeterminacy. For the relations are in the things themselves, not just *between* them. This is, I hope, a deepening of the politically active intersectionality, where all our self-forming issues get entangled.[22] In the interest of the mattering of these matters, we might note that intersectionality was coined by the critical race theorist Kimberly Crenshaw;[23] and already the Combahee River Collective of black women had recognized that you cannot separate race and class and sex issues, coining their own related concept of "simultaneity." If simultaneity in physics formally characterizes the space-time of quantum entanglement, at the social scale of our tangled human relations it conjures the theology of promiscuous incarnations.

THE QUEERNESS OF MATTER

LCS: Yes. "Uncertainty queer" stakes a claim to an elemental open-endedness. But it also regularly pulls up stakes by recognizing that elemental open-endedness can never be fully open, as if anything in the world could be unaffected or indifferent to the limitations of relation. Neither can it be utterly spontaneous since both openness and spontaneity are relational terms. "Open" only has meaning in relation to some limit, does it not? Otherwise it is an empty concept. Spontaneity only has meaning in relation to some determination or expectation.

What makes uncertainty *queer* is therefore always a kind of queer difficulty. In other words, there is never a certainty (she says, definitively!) that is not shot through with wobbly props, the very bread and butter of queer theory. As scientific investigation moves more and more into the slippery edges of uncertainty as a mode of knowing, it comes closer and closer to theological investigation that moves further and further away from dogmatic knowing into the openness and spontaneity of promiscuous divine relation. Scientific investigation and theology are each becoming more queerly aware of their proximity to each other and to the queer matter of (what theology calls) God.

* * *

CK: In Barad's words, "if the indeterminate nature of existence by its nature teeters on the cusp of stability and instability, of determinacy and indeterminacy, of possibility and impossibility, then the dynamic relationality between continuity and discontinuity is crucial to the open-ended becoming of the world which resists acausality as much as determinism."[24] The radical discontinuity of quantum "leaps," whereby what we experience as continuity is created, materializes this paradox as the very basis of matter.

Its "dis/continuity" for her marks queerness, even as she admits to queering queerness itself. "Indeed, given that queer is a radical questioning of identity and binaries, including the nature/culture binary," she argues "that all sorts of seeming impossibilities are indeed possible, including the queerness of causality, matter, space, and time."[25] Queer, then, is not a fixed determinate term; it does not have a stable referential context, which is not, however, to say that it means anything anyone wants it to be. Queer is itself a lively mutating organism, a desiring radical openness, an edgy protean differentiating multiplicity, an agential dis/continuity, an enfolded reiteratively-materializing-promiscuously-inventive spatiotemporality. What if queerness were understood to reside not in the breach of nature/culture, per se, but in the very nature of spacetimemattering? In her illustrations Barad does not stick with elemental physics but solicits an entire "motley crew of queer coworkers—social amoebas, neuronal receptor cells in stingrays, lightning, a phantom species of dinoflagellates, academics (a strange companion species), and atoms among them."[26]

So the very binary of up/down, top/bottom (those pairs, of course, carrying their own sexual patternings) also breaks up. If physics takes us all the way down, it goes at the same time all the way up and out across the universe, undoing any up/down on the way and hardly restricting its *mysterium* to quantum entanglement. A recent cover of *Scientific American* gives the headline version: "Mystery of the Hidden Cosmos: Dark matter may be much weirder than physicists thought."[27] The "dark" does not reference sexual or racial or theological queerness; but it does signal a radical and shifting unknownness before which astrophysics yet—indeed, anew—trembles.

For our present purposes, however, what makes the mystery of matter attractive is not its sheer unknownness but its nonstandard relationality. It is what Barad captures as the "queer performativity of nature." This metaphor of queer performativity arises in her thought as an answer to the culture/nature binary. The opposition of culture to nature has served as the primary tool of heteronormativity. For straight sexual repression and oppression have been perpetrated in the name of the "natural"—a nature meant to be subdued and dominated by culture, and first of all by the cultures of Christianity. In our history, "nature" has been a term of terror for those of us who cannot make themselves fit the dominant cultural norms of nature itself—in gender, in race, and in all of the permutations of embodiment that those categories bear. So women, African Americans, and queers have not fared well in conforming to the terms of that "nature": women and blacks are accused of too much nature, too little culture; but LGBTQ folk are said to perform "unnatural acts," deliberate violations of nature itself. Of course, the charge is always incoherent, as nature is from the heteronormative viewpoint that which must be culturally mastered, productively used, ethically overcome. Yet that culture holds that we all, and so our cultures, have natures. *Are* natural? -ish?

Judith Butler had earlier articulated, building on J. L. Austin, an understanding of speech acts as performative, linking her theory of gender development thus to a linguistic performativity. She famously queered that gender performativity by de*natural*izing it. Barad then reclaims nature by queering it—in performatively Butlerian terms. But she moves beyond the residual anthropocentrism of Butler's work in queer theory (as Butler in a certain measure has also done more recently).[28] So Barad is able to ask: "What if queerness were understood to reside not in the breach of nature/culture, per se, but in the very nature of spacetimemattering?"[29] So identity itself becomes agentially intra-active, at once active and responsive. It is interdependent with everything in its world—that is, it is ontologically relational.

All the way down and, back up and out again. And if we want to use the term nature, it means all of us creatures, human and otherwise, in our agential intra-activities. Here, no spacetime remains for cultural mastery of matter by the mind, let alone of religious sovereignty of the spirit over the flesh. The spacetimemattering dissolves the nature/culture dualism. It does not however thereby yield a homogenous continuum; it does not diminish the difference between bodies, between their so-called natures, between "naturecultures" (Haraway) or spiritual bodies (1 Corinthians 15.33) or promiscuous incarnations. Human cultures, religious and irreligious, cease to count as exceptions that prove the rule of otherwise mindless, meaningless matter. Nature didn't wait for supernatural intervention to produce us; nor for some incoherent evolutionary leap from the merely mechanical up to our cultured minds.

When Karen Barad does subtly engage theology, it is not that of a transcendent supernature but of a Judaism inflected by the Kabbalah,[30] as read in tandem with Walter Benjamin's messianic "now-time,"[31] itself read through Judith Butler—and all set in relation to quantum physics. The *Eyn Sof,* in its virtual infinity, here comes into semantic congruence with the quantum void, in which the very binary of infinity and finitude begins to tremble. Barad attends "to the smallest bits of matter, in order to show that matter has this messianic structure written into its finitude, no matter how small a piece." Indeed, "this is surely true of all material beings, each of which is an enormous entangled multitude."[32]

Interpreting the kabbalistic reading of Genesis through the quantum void of infinite possibilities, she writes: "In the beginning there was an involution, not evolution, a kind of self-touching, such that creation does not simply start with, but materializes with/in a point, an infinitesimal bit of the infinite."[33] The self-touching is a reciprocal touching, a virtually infinite touching, amidst the entangled multitude. In the messianic "flashing up" of the now-time, the entangled differences that historically co-constitute us host the possibilities that may transform us. All of us "queer critters."

The messianic ethic always possibly at play amidst the impossible tangles of the multitude calls across every register of materiality. Perhaps, in the register of theological ethics, this is only to amplify what Emilie Townes (in this volume) called the "background noise" we begin to hear as we unsettle the categories of gender, sex, race, class. And here we add species and materiality, organic and inorganic. Human selves and their sexualities cease to conform to the disconnective norms that reproduce straightness; heterosexism is a case of an abstraction mistaken for actuality, what Whitehead calls "the fallacy of misplaced concreteness."[34] That is the concealment of the concrete multiplicity of relations which make us up, which matter, at any moment—of the touches Whitehead called "prehensions."

Rematerialized, coming back into touch, sexuality then becomes a matter of intra-active responsibility rather than divinely or mechanically predetermined nature. Queerness, far from a new name for another closed set of sexual identities, performs as a natureculture opening into the fabric of embodied interrelations that constitute the creation. A creation made of touches. In his most familiar painting Michelangelo (hardly a model of sexual straightness) hints at the touchingness of the act of creation. Where science and religion touch in Barad that cosmic touchingness does not resemble a hands-on Renaissance sovereign either. We can here only gesture toward the resonance of this infinite set of possibilities with Whitehead's notion of the primordial nature of God, or of its infinity with a creativity not *ex nihilo* but *ex profundis*. Let us, however, leap forward to relate the infinitely touchy quantum field to the more recent work in religion and science of Whitney Bauman.

Bauman's "polyamoury of place" has already articulated for the matter of our interdisciplinarity a captivatingly queer spacetime. His sense of place insists on materialization in a more earthbound spatiality. Its playfully polymorphous perversity manifests that quantum infinity within the—very seriously—ecological entanglements of finite bodies. Its queerness oscillates between science and religion, just there where global warming is rendering the ecological possibilities of earthlings ever more finite. The catastrophe of what he calls "climate weirding" can, he thinks, be faced in a prismatic (not only "green"), and so queerly and multiply, resilient way. The prismatic spectrum enables the multiplicity of disciplines to crisscross the multitude of racial, economic, gender, and sexual identifications mindfully—when they are honest—touching each other, endlessly and sometimes terrifyingly.

Climate catastrophe demonstrates our ontological relationality as much as does quantum queerness. *Tremendum et fascinans.* And the point is that when science points us to the multiple levels of the intra-activity of human and nonhuman, we cannot extract ourselves from nature. From world. From the vibrant mattering of the universe. Promiscuously incarnated, polyamorously placed.

A QUEER POSSIBILITY OF THEOLOGY AND SCIENCE IN RELATION

LCS: We have here the possibility of queer theory—as a mode of criticism and of pleasure—that can, weirdly, bring theology and science into a different dialogic relation than they've had in the past. To the extent that theology seeks poetic and explicative modes that attend not only to inherited texts, traditions, and moral claims (about love of neighbor, say) but that seek to engage and even attempt some description of worldly relation with divinity, queer theory drives theology closer to the edges of scientific inquiry. It helps attune theology to its tendencies to normalize and domesticate both god and world and pushes theology to be more suspicious of its gendered and racialized fears of indecency. This affective turn to indecency occurs in no small part because queer theory tends away from any attempt at sanitization of bodies, whether atomic, mammalian, or planetary. So for Christian theologians, as we said in the beginning of this dialogue, the imperative toward incarnation lends queer theory a strong voice in the theological effort to take that materiality of divinity both seriously and adventurously.

The move to critically investigate decency in contemporary queer theology brings religious discourse closer to the cutting edges of scientific inquiry than we've found to be the case before. Our lay understanding of quantum science, biology, botany, and the other sciences as they push themselves into deeper awareness of interconnection and deep relation (which naturally illuminates earlier dogmatisms and epistemological hubris) begins to seem more and more compatible with what is going on in theology. The edges of uncertainty overlap, the doctrines of method prove porous to each other, the line between faith and habit becomes blurry.

But let us not make the mistake of suggesting that scientific pursuits and theological pursuits are habits of the same type. No—what we see from the theological side is that our differences are becoming more compatible. Both are becoming more interested, attracted, fascinated, by uncertainty—which requires different approaches to doctrines and methods on both sides. Both are becoming more interested, attracted, fascinated, by the constitutive powers and affects of relation/s, which require different approaches to doctrines and methods. Both are becoming more aware of the tremendous dangers, the catastrophic potentials, of our ability to distort understanding and so wreak havoc and pain in our wake. Tremendum indeed. The stakes are high, we have some compatible points of discussion, many places we touch! Queer theory is not an answer, not a result, not a doctrine or a law. It is merely a lens, a lens onto a world that is always already looking back at us.

Because . . .

* * *

CK: "Nothing in this world is indifferent to us."

I may now out the author of that line: none other than Pope Francis, in *Laudato Si'*. His encyclical is one great protest against indifference—against the systemic disconnections that close our ears to "the cry of the poor and the cry of the planet," against the "throwaway culture which affects the excluded just as it quickly reduces things to rubbish" (20, 22), that is betraying that "common good" that is the climate of all of our lives.[35] He insists upon an "integral ecology" engaging faith in natural science while critical of technocracy. As the single most influential religious voice on the earth, he appeals to all who will listen. While certain Catholic politicians were immediately asking him to leave science to the scientists, they of course are the science deniers. He is educated and certified in chemistry. But—in queerness? There is nothing heartening about the Pope's formal positions on sex, gender, or reproduction. We just recall that he only got so much of the attention of liberal Catholics and others of us long discouraged by papal performances because of his statement on an airplane early in his papacy: "Who am I to judge?" (In terms of sexual politics, we cannot prove that this opening allowed the Irish to vote as they did, or Catholic Justice Kennedy to cast the deciding vote on same sex marriage, soon thereafter.) But as to the identity-entangling spacetimemattering of us all, which is to say, the deeper queerness of ontological inseparability: he is right there. In his theistic translation, the infinite touching reads thus: "Soil, water, mountains: everything is, as it were, a caress of God" (84).

Pope Francis might say amen to Barad's cosmic metaphor: "a delicate tissue of ethicality runs through the marrow of being."[36] That ethicality names the alternative to cosmic indifference. But the strict opposite of indifference is difference, which makes ethics possible. Ethicality is response to difference, calling for responsibility. Difference—the particularity of existence—matters. If everything in the world exists only in direct or indirect relation to everything else, it all forms a tissue of endless difference. It makes everything different from everything else. But difference is relationship: and it is in the complex of our differences, not in the nihilistic unity of indifference, that we have our 21st century chance as a species. That possibility—that we might make the evolutionary leap to a sane and sustainable future—keeps us right on the edge of the unknown, doesn't it? The *mysterium tremendum* of our planetary future has us, or should have us, trembling at the edge of the 6th great extinction event: what Elizabeth Kolbert calls "an unnatural history," the earth of spreading fires and droughts, melting glaciers and icecaps, rising seas and floods. Not that climate change supersedes our other issues, our differences of natureculture, of material and historical reality. Rather, it interlaces sexuality with economics, race, disability, in the mental, physical, spiritual ecology of the earth. And so we may also be awakening to the *mysterium fascinans* of our agential, erotic, responsive entanglement. Such responsiveness may intensify a sense of responsibility beyond moralism, or, as Barad puts it, in a point of view queerly poised right between science and religion: "For me ethics is not a concern we add to the questions of matter, but rather is the very nature of what it means to matter."[37]

Perhaps a new ethical intensity begins to bubble up in resistance to aggressive modes of both science-denialism and spiritual indifference. Might it draw all of us queerly entangled, climate-dependent, earth-dwelling critters, trembling and attracted, into new alliance?

WORKS CITED

Althaus-Reid, Marcella. *Indecent Theology*. New York: Routledge Press, 2000.

Bacon, Sir Francis. "Book One" in *The New Organon.* Cambridge: Cambridge University Press, 2000.

Barad, Karen. *Meeting the Universe Halfway*. Durham, NC: Duke University Press, 2007.

———. "Nature's Queer Performativity." *Kvinder, Køn & Forskning* 12 (2012), 25–53. Also published in *Qui Parle: Humanities and Social Sciences* 19 no. 2 (2011): 121–158.

———. "What Flashes Up: Theological-Political-Scientific Fragments." In *Entangled Worlds: Religion, Science and New Materialisms.* Edited by Catherine Keller and Mary-Jane Rubenstein. New York: Fordham University Press, 2017.

Benjamin, Walter. "Theses on the Philosophy of History." In *Illuminations: Essays and Reflections*. Schocken: NY: 1968 [orig published in German 1955].

Butler, Judith. *Bodies That Matter: On the Discursive Limits of Sex*. New York: Routledge, 1993.

Chen, Mel Y. *Animacies: Biopolitics, Racial Mattering, and Queer Affect.* Durham, NC: Duke University Press, 2012.

Collins, Patricia Hill and Sirma Bilge. *Intersectionality*. Cambridge: Polity Press, 2016.

Davies, Paul. *God and the New Physics*. New York: Simon and Schuster, 1983.

Dobrescu, Bogdan A. and Don Lincoln. "Mystery of the Hidden Cosmos." *Scientific American* 313 (2015) 32–39. doi:10.1038/scientificamerican0715–32.

Faber, Roland, ed. *Butler on Whitehead: On the Occasion*. Lanham, MD: Lexington Books, 2012.

Francis. *Laudato Si'*. Encyclical Letter of the Holy Father Francis on Care for Our Common Home. Vatican Website. June 2015. http://w2.vatican.va/content/francesco/en/encyclicals/documents/papa-francesco_20150524_enciclica-laudato-si.html

Keller, Catherine. *Cloud of the Impossible: Negative Theology and Planetary Entanglement.* New York: Fordham Press, 2015.

Keller, Catherine and Mary-Jane Rubenstein eds. *Entangled Worlds: Religion, Science, and the New Materialisms.* New York: Fordham University Press, 2017.

Malin, Shimon. *Nature Loves to Hide: Quantum Physics and Reality, a Western Perspective.* New York: Oxford University Press, 2003.

McClintock, Anne. *Imperial Leather: Race, Gender, and Sexuality in the Colonial Context*. New York: Routledge, 1995.

Muñoz, José Esteban. *Disidentifications: Queers of Color and the Performance of Politics*. Minneapolis, MN: University of Minnesota Press, 1999.

Otto, Rudolf. *The Idea of the Holy.* Oxford: Oxford University Press, 1958.

Puar, Jasbir. *Terrorist Assemblages: Homonationalism in Queer Times.* Durham, NC: Duke University Press, 2007.

Schneider, Laurel C. "Queer Theory." In *Handbook of Postmodern Biblical Interpretation,* edited by A. K. M. Adam. St. Louis: Chalice, 2000.

———. "What Race Is Your Sex?" *Queer Religion: LGBT Movements and Queering Religion Volume II*. Edited by Donald L. Boisvert and Jay Emerson Johnson. Santa Barbara, CA: Praeger, 2012. 125–42.

Sedgwick, Eve Kosofsky. *Tendencies.* Durham, NC: Duke University Press, 1993.

Whitehead, Alfred North. *Science and the Modern World*. New York: The Free Press, 1925.

NOTES

1. Laurel C. Schneider, "Queer Theory" in *Handbook of Postmodern Biblical Interpretation,* ed. A. K. M. Adam (St. Louis: Chalice, 2000).

2. For a more thorough discussion of the sex-race co-constitution in relation to queer theory, see Laurel C. Schneider, "What Race Is Your Sex?" in *Queer Religion: LGBT Movements and Queering Religion, Volume II*, ed. Donald L. Boisvert and Jay Emerson Johnson, 125–42 (Santa Barbara, CA: Praeger, 2012).

3. Sir Francis Bacon posited four "idols of the mind" that beset scientific thinking. See Book One in *The New Organon* (Cambridge: Cambridge University Press, 2000).

4. See Jasbir Puar,, *Terrorist Assemblages: Homonationalism in Queer Times* (Durham, NC: Duke University Press, 2007) or Anne McClintock, *Imperial Leather: Race, Gender, and Sexuality in the Colonial Context* (New York: Routledge, 1995) et al. on nationalized/racialized sexualities.

5. Marcella Althaus-Reid, *Indecent Theology* (New York: Routledge Press, 2000).

6. Mel Y. Chen, *Animacies: Biopolitics, Racial Mattering, and Queer Affect* (Durham NC: Duke University Press, 2012).

7. I recommend here Catherine Keller's recent work in negative theology that negates mere negativity: *Cloud of the Impossible: Negative Theology and Planetary Entanglement* (New York: Fordham Press, 2015).

8. See Karen Barad, "Nature's Queer Performativity" in *Kvinder, Køn & Forskning* 12 (2012).

9. Rudolf Otto, *The Idea of the Holy* (Oxford: Oxford University Press) 1958.

10. José Esteban Muñoz, *Disidentifications: Queers of Color and the Performance of Politics* (Minneapolis, MN: University of Minnesota Press, 1999).

11. See Emilie Townes's contribution to this volume, "If You Quar It You Can Change It: Changing the Boxes that Bind Us."

12. Eve Kosofsky Sedgwick, *Tendencies* (Durham, NC: Duke University Press, 1993), 9.

13. Werner Heisenberg, in *The Physicist's Conception of Nature* (1958), cited in Paul Davies, *God and the New Physics* (New York: Simon and Schuster, 1983), 80.

14. Shimon Malin, *Nature Loves to Hide: Quantum Physics and Reality, a Western Perspective* (New York: Oxford University Press, 2003), 80.

15. Karen Barad, *Meeting the Universe Halfway* (Durham, NC: Duke University Press, 2007), 309.

16. Ibid., 67.

17. Ibid., 66.

18. Laurel Schneider, "Promiscuous Incarnation," in *The Embrace of Eros: Bodies, Desires, and Sexuality in Christianity*, ed. by Margaret Kamitsuka, 231–246 (Minneapolis: Fortress Press, 2010).

19. Karen Barad, "Nature's Queer Performativity," 44.

20. Ibid., 29.

21. Townes, "If You Quare It You Can Change It," in this volume, 53.

22. For a helpful discussion of intersectionality, see 23 and 29 (internal citation to introduction)

23. For an update of the concept of intersectionality, see Patricia Hill Collins and Sirma Bilge, *Intersectionality* (Cambridge: Polity Press), 2016.

24. Barad, *Meeting the Universe Halfway*, 182.

25. Barad, "Nature's Queer Performativity," 29.

26. Ibid.

27. Bogdan A. Dobrescu and Don Lincoln, "Mystery of the Hidden Cosmos," *Scientific American* 313 (2015) 32–39. doi:10.1038/scientificamerican0715-32.

28. See, for example, the recent volume edited by Roland Faber et al., *Butler on Whitehead: On the Occasion* (Lanham, MD: Lexington Books, 2012), or even Judith Butler's own *Bodies That Matter: On the Discursive Limits of Sex* (New York: Routledge, 1993). Note also Keller, "Unsaying and Undoing: Judith Butler and the Ethics of Relational Ontology," in *Cloud of the Impossible,* 215–238.

29. Barad, "Nature's Queer Performativity," 53.

30. See Fern Feldman's chapter in this volume.

31. Walter Benjamin, "Theses on the Philosophy of History," *Illuminations: Essays and Reflections* (New York: Schocken, 1968 [orig, published in German 1955]), 261.

32. See Barad's essay, "What Flashes Up: Theological-Political-Scientific Fragments," in the volume edited by myself [Catherine Keller] and Mary-Jane Rubenstein, *Entangled Worlds: Religion, Science, and the New Materialisms* (New York: Fordham University Press, 2017) 70, 73. This volume is a collection of essays delivered at the 2014 Drew Transdisciplinary Theological Colloquium.

33. Ibid., 52.

34. The "fallacy of misplaced concreteness" is "merely the accidental error of mistaking the abstract for the concrete." Alfred North Whitehead, *Science and the Modern World* (New York: The Free Press, 1925).

35. Francis, *Laudato Si'*. Encyclical Letter of the Holy Father Francis on Care for Our Common Home. (Vatican website, June 2015), http://w2.vatican.va/content/francesco/en/encyclicals/documents/papa-francesco_20150524_enciclica-laudato-si.html.

36. Barad, *Meeting the Universe Halfway,* 392.

37. Barad, "What Flashes Up," 70.

Chapter Eight

Queering the Dissident Body

Race, Sex, and Disability in Rabbinic Blessings on Bodily Difference

Julia Watts Belser

In rabbinic Jewish practice, observant Jews are instructed to recite a blessing that praises God for creating diverse creatures, a blessing that has been both praised and pilloried by contemporary Jewish writers.[1] Meant to be spoken when a pious Jew confronts an unusual or uncanny individual, the recitation of *meshane habriyot* serves as a striking means of negotiating corporeal difference. The blessing marks out a number of individuals as objects of fascination: people with dark skin, disability, and other forms of stigmatized bodily difference become an occasion for recognizing the diversity of human kinds. Reflecting on the ethical imperatives that underpin this practice, Louis Newman argues that the blessing requires a practitioner to overcome the common human tendency "to look away" from that which we find repellant, to "look past the superficial features that we find disturbing; instead, we are to see what God sees, another human being bearing the divine image."[2] Lauren Tuchman, by contrast, argues that while the blessing is often praised for its call to affirm and appreciate human diversity, the practice is experienced quite differently by those on the receiving end of the blessing. As a blind woman who has been the frequent object of fellow Jews' pious recitations, Tuchman argues that instead of "honoring the beauty within our non-normative bodies," the blessing places individuals with disabilities and other visible differences "firmly within the category of the 'other.'"[3]

In this chapter, I analyze the blessing *meshane habriyot* through the lens of queer theory, critical race theory, and feminist disability studies. I examine its articulation in the classic rabbinic sources, from its first textual emergence in second century Palestine to its elaboration and development in late antique Babylonia. The blessing, I assert, serves to make visible the geographies and taxonomies of otherness, affirming natural diversity even as it strives to shore up the boundaries of the normative human body. While the blessing itself aims to locate "strangeness" as a property of certain human bodies, its recitation reveals the way in which otherness is not a natural difference, but a fashioned one. Thinking with Judith Butler, I argue, the rabbinic blessing reveals what we might call *the performativity of deviance*: a claim that otherness is best understood not as "a substantial thing or a static cultural marker, but rather as an incessant and repeated action."[4] Rabbinic debates over the blessing's proper practice reveal the fundamental instability of racial and corporeal categories, showing up cracks and fissures within the conceptual apparatus that maps out deviance.[5] But while such moments of categorical instability might lure the reader toward dreams of queer subversion, the embodied rabbinic practice of blessing works to produce, constrain, and concretize deviance as a discrete category, rooted in the organic physicality of certain human kinds. Affirming Butler's emphasis that gender is "a ritual reiterated under and through constraint," I argue that the rabbinic practice constitutes both the normative self and the dissident other by governing the gaze, by regulating and ritualizing the act of assessing and blessing the strange.[6]

SPECTACULAR SKINS: COLOR AND THE PRODUCTION OF DIFFERENCE IN *MESHANE HABRIYOT*

In rabbinic Jewish tradition, blessings punctuate virtually every moment of the day: from blessings recited upon awakening to blessings recited after using the toilet, blessings before and after meals, as well as blessings that acknowledge the morning light and the twilight. The blessing *meshane habriyot*, which praises God for creating diverse creatures, appears as part of a series of blessings recited after seeing wondrous or unusual elements of divine creation. In rabbinic thought, such recitations are a religious obligation, not a spontaneous practice—a matter of rabbinic law, not individual desire. Rabbinic texts establish the conditions under which blessings are to be recited and the specific words used when speaking the blessing. The call to bless diverse creatures first appears in the Tosefta, a non-canonical second century Palestinian rabbinic text.[7] Both the Palestinian and Babylonian Talmud, key rabbinic texts whose redaction is usually dated to the fifth and seventh centuries respectively, discuss the blessing *meshane habriyot* and regard it as a religious obligation. Let us begin by examining Tosefta Brakhot 6:3:[8]

> One who sees a black person (ha-kushi),[9]
> a "giḥor" (usually translated as "red," but occasionally as "white"),
> a "lavqan" (usually translated as "white," but occasionally as "red"),
> a person with a hunchback/who is very tall (ha-qipeaḥ),[10]
> a person of short stature (ha-nanas; often translated as "midget" or "dwarf"),

[a person with dropsy (hidraniqos)],[11]
[a deaf person (ha-ḥeresh),
a person who is not mentally competent (ha-shoteh),
or a person who is drunk (ha-shikor)],[12]
he says, "Blessed be the One who varies (meshane) the creatures."

{One who sees] an amputee (ha-qiteʿa),[13]
a lame person (ha-ḥiger),
a blind person (ha-suma),
a person struck with boils (mokhei-shiḥin),
[a person whose head is flat (pittuyei rʾosh),
a person with a white-spot eruption (ha-bahaqa,nin)],[14]
He says, "Blessed be the true judge."

The Tosefta's tradition begins with three categories associated with distinctive skin pigmentation: blackness, redness, and whiteness. The first term, "the kushi," derives from the biblical place-name of Cush, a land that lies south of Egypt.[15] While the Hebrew Bible frequently uses *kushi* to refer to the Nubian people, who lived along the Nile River in modern-day Sudan, biblical genealogy also identifies Cush as a son of Ham, the cursed son of Noah.[16] Beyond these specific designations, Abraham Melamed argues that the biblical writers use Cush as a metaphor for "a distant and different place, at the ends of the earth."[17] Though the term *kushi* develops a clear negative valence in later Jewish literature, Melamed argues that biblical narratives largely lack this moralistic quality. In rabbinic texts, by contrast, the term *kushi* takes on a decidedly negative meaning—a transformation that mirrors the shifting color discourse of Greco-Roman culture, which likewise turned away from the more neutral conception of blackness in ancient Greek society.[18] In the rabbinic Jewish imagination, the blackness and foreignness of the *kushi* become overwhelmingly read as markers of moral deviance.[19]

If the designation *kushi* has a clear (if still quite complex) referent in rabbinic discourse, the Tosefta's other two color terms are more ambiguous. Reading these texts through the prism of queer theory and critical race theory makes visible a startling instability at the heart of these categories, a contestation over meaning that seems to unravel the very practice that rabbinic texts are striving to establish. Consider a debate that appears in another tractate of the Babylonian Talmud, which discusses the meaning of the terms *giḥor* and *lavqan,* asserting that what here means "red" is actually "white" and vice versa. Amidst a discussion of animal bodies, Bavli Bekhorot 45b asserts:

Kushi is black.
Giḥor is white.
Lavqan is red.

Is this so?
But there was one who said, "Who wants *lavqanim*?"
And they were found to be white!

Such linguistic uncertainty is not uncommon in rabbinic discourse. Rabbinic narratives and rabbinic legal queries often turn on the precise meaning of particular words. An oft-told tale about two rabbis who are baffled by particular mishnaic terminology imagines the rabbis "discovering" the meaning of the terms by watching the maidservant of Rabbi Judah the Patriarch, who uses those very words in ordinary speech.[20] Reading this tale, Dina Stein observes the way it overturns the reader's expectations regarding knowledge and virtuosity. It is, she argues, "the destabilized institutional order" that "produces the newly acquired linguistic knowledge."[21] In our text, as well, the rabbis secure the meaning of the term by turning to "ordinary" practice. The shepherds hawking their lambs on the street have retained a knowledge the rabbis seem to have lost.

In Bavli Bekhorot, rabbinic debate makes visible the constructed nature of these color categories, which often function—in humans, though not in animals—as significant markers of somatic difference.[22] These texts expose a categorical instability within the Tosefta's blessing schema, an instability that evokes something of Michael Omi and Howard Winant's influential assessment of race as "an unstable, decentered complex of social meanings."[23] In translating these texts, I have sought to avoid the illusion of a single stable meaning. In ascribing meaning to the Tosefta's terms, I could decide to give determinative weight to later rabbinic interpretive tradition, or to the commentary traditions that have shaped the reception of this text for Jewish readers. Such readings are common in religious contexts, though not in the secular academic circles in which I have trained. Following my scholarly instincts, I might turn to linguistic etymology to confirm the meaning of *lavqan;* the Hebrew is derived from the Greek words *leukoi* and *leuchan*, both of which mean "white." But either move risks papering over the queer fact of the rabbis' own uncertainty. Thus, I ask you to linger with me in the question, to tarry with the strangeness of this ambiguity. Even as rabbinic tradition aims to establish a practice built on the observation of visible skin difference, it confesses insecurity about the very colors it uses to signify otherness. It knows it must bless bodies that are white and red, but it is not sure which are which.

Let us, for the moment, accept as a given the premise that *lavqan* denotes "white." Even if this is true, what does it mean? *Lavqan* is often translated as "albino," a concept that has the benefit of aligning neatly with modern reader's sensibility about the conditions that might render white skin strange. Historians of disability have, in recent years, become particularly attuned to the problem of ascribing contemporary diagnoses to ancient subjects, observing the way that such scholarly interventions can reify and naturalize modern medical science as though it were a neutral, objective arbiter of bodily truth.[24] Even beyond the problems of retrospective diagnosis, I argue that a smooth translation of *lavqan* as albino has other troubling discursive effects. With this move, "ordinary" whiteness is constructed as an unmarked category, while only "extraordinary" whiteness serves as an occasion for curiosity. The Black body, by contrast, is uniformly read as unusual. Blackness is rendered as an anomaly, constructed in parallel not to quotidian whiteness but to albinism—a condition most moderns would reflexively identify as a medical pathology.[25]

Parsing the production of otherness within the Toseftan tradition reveals a complex interrelationship between discourses of race and corporeal deviance. Consider the way in which modern commentators have attempted to make sense of the elusive *giḥor*. One mode of reading takes it as an "ethnic" marker, following a conception of Native Americans as "red" peoples.[26] In his recent commentary on the Tosefta, Eliyahu Gurevich rejects such a reading on the grounds that "there was no possibility in talmudic times for an American Indian to appear in the Middle East."[27] Even as it works to deny the applicability of the racial category, Gurevich's gloss serves to fix ethnic identity in place. "Red skin," read racially, becomes a property of a specific group of people, defined as both culturally and geographically unknowable to the rabbis.

Rejecting *giḥor* as a marker of ethnicity, Gurevich reads redness as a different kind of somatic anomaly, suggesting that our text is directed toward "a redheaded person, [who] specifically has red spotted skin with a lot of red freckles." Yet even as he proposes this alternative, Gurevich argues that it poses a conceptual difficulty, for "many Greeks," he maintains, "had red hair and freckles, at least mildly."[28] It is not my interest to enter into a discussion of the material likelihood that red hair might have been a common sight for the rabbis, but rather to make apparent the assumptions on which this argument is based: that familiarity would undermine the Tosefta's investment in the blessing. To be properly unusual, the *giḥor* must be other than the expected Jew or Greek.[29] Together with the *kushi* and the *lavqan,* the *giḥor* reveals the way in which skin difference signals either a betrayal of the expected nature of a familiar ethnic category, or a racialized foreignness that frames the anomalous body as a product of queer geography: the presence of a body out of place.

"NATURALIZING" THE DEVIANT BODY: DISCOURSES OF RACE AND DISABILITY

Analyzing these rabbinic traditions, I use the term "race" deliberately, recognizing with Geraldine Heng that race is a term that "continues to bear witness to important strategic, epistemological, and political commitments not adequately served by the invocation of categories" like ethnic difference or bodily otherness.[30] While there is substantial overlap between the concepts of ethnicity and race, Cornell and Hartmann argue that race is category defined "by virtue of perceived common physical characteristics that are held to be inherent."[31] This perception of biophysical truth is both a defining quality of racial thinking. Such biophysical claims are strongly critiqued by modern science and by critical race theorists. Race, they contend, is not dictated by biology; it is a human invention. Such choices are never neutral. "The history of race," Cornell and Hartmann emphasize, "is a history of moral judgments, a division of the world into more or less worthy categories of persons."[32] This assertion of biological difference, which undergirds the logic of racial formation, differs from ethnicity in its sticky insistence that racial designations reveal inherent biophysical differences between peoples, that the categorical unworthiness borne by certain races is rooted in the truth of the body.

A word about my choice to attend to rabbinic thinking about race: The analysis of race in premodern cultures has often been foreclosed by the claim that "scientific racism" is an invention of the European Enlightenment. In deploying the category of race to probe rabbinic conceptions of bodily difference, I do not mean to imply a seamless intellectual trajectory between the Tosefta and the taxonomy of modern eugenic racial sciences. Nor do I believe that such imperatives drive rabbinic assessments of biophysical difference. Yet I affirm Heng's recognition that "*not* to use the term race would be to sustain the reproduction of a certain kind of past," a choice to "de-stigmatize the impacts and consequences" of particular texts, laws, and practices in premodern culture.[33] Following Cord J. Whitaker, I ask "not whether, but how" late antique notions of race operate.[34] In this chapter, I heed the call of scholars working at the intersection of race and disability studies to take seriously the way in which racial formation is fueled by discourses of disablement. Sharon Snyder and David Mitchell have argued that the eugenic sciences of the post-Enlightenment Atlantic leveraged disability in order to make the case for racial inferiority. "Disability," they emphasize, "came to be construed as a socially dehumanizing construct in tandem with theories of racial degeneracy."[35]

The rhetorical deployment of disability to justify inequality holds a significant if oft overlooked place in American history, undergirding the logics of African American slavery while also driving opposition to women's suffrage, civil rights, and immigration. As Douglas C. Baynton has shown, "the *concept* of disability has been used to justify discrimination against other groups by attributing disability to them."[36] Medicalized notions of hysteria, irrationality, and physical weakness have often been used to challenge women's equality, while arguments for racial inequality and restrictions on immigration have often rested on notions of physical difference, intellectual inequality, and "feeblemindedness."[37] While scholars and activists have often sought to distance women, immigrants, and communities of color from associations with disability, Baynton observes that oppressed groups have almost never argued against the notion that "disability is an adequate justification for social and political inequality."[38] Such a choice, Nirmela Erevelles argues often "unwittingly reaffirms an imagined biological wholeness (normativity) that was instrumental in the propagation of the same oppressive ideologies." She calls, instead, for a more nuanced recognition of the way in which "race and disability are imbricated in the construction of the pathological Other," as well as critical attention to the "historical contexts and structural conditions within which the identity categories of race and disability intersect."[39]

Critical race scholars have challenged the salience of racial taxonomies, emphasizing that racial categories are constructed differences, not biological facts. Disability studies theory has likewise challenged the notion that disability is rooted in bio-physical impairment, asserting that disability is created as much by social and cultural barriers, as it is by organic human difference. Disablement, disability studies theorists claim, is not a straightforward fact of the body, but a product of social stigma, architectural inaccessibility, and the historically situated demands of particular economic and social ideologies.[40] Yet despite such critical interventions, popular discourse continues to insist that racial and disability difference are lodged in the flesh. How shall we understand this fascination with the organic physicality of difference, this desire to root race and disability in the visible body? Etrienne Balibar argues that proponents of racism have often sought to "mimic scientific discursivity by basing themselves upon visible 'evidence'"—on the "bodily stigmata" of racial difference.[41] Early anthropological science, Fatimah Tobing Rony asserts, likewise construed racialized bodies as sites of disablement, a perception made vivid in the journals of Félix-Louis Regnault, who described the gatherings of Indian pilgrims he observed as "real museums of the pathological," gatherings of "lepers, those with elephantiasis, microcephalics, and the deformed of all kinds."[42] The discourse of defectiveness and pathology, Eli Clare maintains, has long been used "to justify the practices of enslavement, imprisonment, institutionalization, and state violence." Such claims, he argues, "fortify white supremacy by leveraging ableism."[43]

PERCEIVING PERVERSION: DISABILITY, RACE, AND SEXUAL DEVIANCE

Though our Tosefta text begins with skin difference, it soon turns to terrain more conventionally associated with disability. Its next two categories call the witness to bless God upon seeing a person whose body has a non-normative shape: a person with a hunchback/a person who is very tall [*qipeaḥ*] and a person of short stature [*nanas*], usually translated as "a dwarf."[44] While the *qipeaḥ* is not a particularly well attested term in rabbinic literature, the *nanas* occupies a more significant place in the rabbinic imagination. Rabbinic tradition describes Nebuchadnezzar, the Babylonian king who conquered Judah, as a *nanas,* using his small stature as a derogatory epithet that underscores the indignity of Jerusalem's fall. Glossing the beginning of Psalm 137 that describes the captive Israelites weeping by the river of Babylon and hanging up their harps, a Palestinian midrash imagines the exiles responding to Nebuchadnezzar's command to perform: "Is it not enough of a torment for us that by our sins we destroyed His temple? Must we now strike our harps for this dwarf?"[45] Bavli Moed Katan 18a imagines another oppressor of the Jewish people as a man of short stature:

> Avitul the Scribe said in the name of Rav:
> The Pharaoh of Moses' day was a cubit [in height].
> His beard was a cubit and his penis was a cubit and a half,
> in order to fulfill that which is said:
> *God appoints the lowest of men over [the kingdom of man.]* (Daniel 4:14)

In this tradition, Pharaoh's physical difference becomes a visible indicator of his impaired virtue; the smallness of his stature becomes a metaphor for his moral character. But Pharaoh's body is not simply a normative male form in miniature. Instead, the talmudic tradition emphasizes the disproportionality of his parts. Against the smallness of his body, Pharaoh's beard and his phallus become gargantuan.[46]

The rabbinic dwarf is a decidedly queer figure, a person whose striking bodily difference is often associated with deviant sexuality. While the tradition of Pharaoh's outsize phallus locates sexual deviance in the physical body of the small-statured man, other rabbinic traditions fashion the disabled body as a visible sign of transgressive sexuality. In a move reminiscent of racialized fantasies that intertwine sexual perversity with ethnic otherness, Genesis Rabbah 37:5 imagines the nation of the Caphtorim mentioned in Genesis 10:14 as a nation of dwarves, a nation that came about because their ancestral progenitors, the Pathrusim and the Casluhim, "set up public bazaars in which they stole each others' wives." This rabbinic tradition imagines short stature as the property of an entire people, offering an origin story that links the creation of the nation with sexual permissiveness and perversion. Similar imaginaries also shape rabbinic perceptions of the children of Ham, linking blackness with sexual perversion. Bavli Sanhedrin 108b argues that Noah's son Ham was one of three creatures who engaged in sexual relations in the ark, a sin for which Ham "was smitten (*laqah*) in his skin"—a tradition that the Palestinian Talmud reads explicitly as becoming dark.[47] Such associations were widespread in the late antique Roman world. Ania Loomba argues that the association between blackness and lechery appears as early as second century CE Roman sources, which associate black people with heat and unrestrained sexual passions.[48]

If discourses of sexual perversity are often scripted onto the nation, they are also used to tie disablement to the sexual sin of specific individuals. Several Babylonian rabbinic traditions explain the birth of children with physical and sensory disabilities as indicative of the mother's sexual and dietary habits or, alternately, with the immodest sexual behavior of the parents.[49] Consider the tradition of Rabbi Yoḥanan ben Dahavei in Bavli Nedarim 20a:

> Rabbi Yoḥanan ben Dahavei said:
> The ministering angels told me four things—
> Why do lame ones come to be?
> Because they overturn their table.
> Why do mute ones come to be?
> Because they kiss that place.
> Why do deaf ones come to be?
> Because they speak during sex.
> Why do blind ones come to be?
> Because they look at that place.

This text marshals the cultural undesirability of disability to bolster rabbinic sexual norms, offering what Michael Satlow has called a "eugenic argument" for proper sexual behavior.[50] Using figurative language that evokes transgressive sexual practice, the Babylonian Talmud offers an etiology of disability that links disablement to parental perversion.[51] "Lame ones" come into the world because the couple "invert" their sexual positions, a phrase that might indicate turning the woman over during intercourse or a mode of intercourse in which the woman is on top.[52] Rabbi Yoḥanan ben Dahavei's claim is challenged by the Babylonian Talmud, in favor of an argument that "whatever a man desires to do with his wife, he can do"—a passage that, in Daniel Boyarin's reading, ultimately privileges rabbinic authority, over and against the claims of women or angels.[53] Even as I affirm, with Boyarin, the way such rabbinic debates are "carried out across the discursive bodies of women," I draw our attention in another direction: the way this rejected and yet still remembered tradition positions the disabled body as a visible marker of "queer" sex.[54] The disabled body becomes visible evidence of past sexual deviance, a way of imagining the living and lingering after-effects of perverse pleasures.

CONGENITAL DIFFERENCE OR ACCIDENTS OF AFFLICTION: CATEGORIZING BODILY DIFFERENCE

While the rabbinic texts just discussed grapple with congenital disability, rabbinic blessing practices also encompass disabilities acquired over the course of a life. The Tosefta divides spectacular bodies into two different categories: the first provides an occasion to praise God for creating diverse creatures, while the other calls for blessing God as "the true judge." While the Tosefta never elaborates on the conceptual basis of this distinction, the Palestinian Talmud is more forthcoming:

> [One who sees] an amputee (ha-qiteʿa),
> a blind person (ha-suma),
> a person struck with boils (mokhei-shiḥin),
> he says, "Blessed be the true judge."
>
> This teaching applies if he was whole (shlamim) and changed (nishtenu).
> But if he was so since he came from his mother's belly,
> He says, "Blessed be the One who varies (meshane) the creatures."[55]

Not only does the Yerushlami's tradition assert a categorical difference between what moderns call congenital and acquired disabilities, it affords that difference theological significance. "If one is born without disabilities and they later develop," Judith Abrams concludes, "then the disabilities are a judgement from God. Those born with disabilities, however, are simply among God's varied creatures."[56] Through this blessing practice, rabbinic tradition interprets certain human conditions as evidence of divine creativity and others as an expression of divine judgment.

But should we indeed read this tradition as a claim that the rabbis viewed acquired disability as an explicit act of divine punishment? I think not. The recitation, "Blessed be the true Judge," is laced with negative associations in rabbinic thought. Mishnah Brakhot 9:2 instructs practitioners to recite "Blessed be the true judge," when hearing bad tidings. The words remain, to this day, the traditional Jewish response to learning of a death. While some commentators have used this parallel to argue that disability is a state akin to death, such claims seem to tell us more about the commentator's fears of disability than the rabbis' own views.[57] With regard to divine punishment, it would seem a gross mischaracterization to claim that reciting "Blessed be the true judge" after hearing of a death is meant to indict the sinfulness of the deceased. Christine Hayes argues that rather than making "a robust metaphysical claim" about divine judgment, the blessing serves "a psychological and spiritual purpose, shoring up faith in God at a time when that faith is most severely tested." When faced with a situation that appears to call divine justice into question, the speaker rejects the conclusion that God "is a negligent or corrupt judge." Even as the blessing affirms divine righteousness, Hayes concludes, it "seems to mask a deep anxiety about the justice of the verdict."[58] While this tradition does position acquired disability as an experience of misfortune, the rabbinic blessing need not be read as an indictment of the individual who acquires a disability. More significant for my own argument is the way this tradition frames disability as a provocation—a site that challenges the viewer, that demands a response.

STAGING THE RABBINIC FREAK SHOW: BLESSING AND THE PERFORMATIVITY OF DEVIANCE

Rabbinic visual blessings, Rachel Neis has argued, "seek to ritualize, guide, and shape affect for and experience of each visual object."[59] This attention to the experience and meaning of sight reflects a rabbinic perception that vision is an affective experience, that it shapes the consciousness of the one who sees. In the ancient Mediterranean and Near Eastern worlds, Neis maintains, "to see something was to be spiritually and physically affected by it."[60] The rabbinic blessing thus aims to shape the one who sees, to ritualize the act of looking at the strange so that it becomes properly formative. The rabbinic blessing is staged as a three part relationship between the speaker ("the one who sees"), the person who prompts the blessing ("the creature"), and God, who is acclaimed through the blessing as either the One who "varies the creatures" or is their "true judge." While the reader might be tempted to position the Black person or the amputee as the object of the blessing, this is not entirely correct. The blessing itself is directed toward God, not the object of the speaker's gaze. The one who is seen appears in our triad merely as the occasion for blessing; their function in this exchange is to provide visual confirmation of a truth the blesser sees and affirms about God. The blessing thus positions "the creature" at the center of the visual dynamics that make the blessing possible, even as it refuses the creature's capacity to participate in the dialogical exchange at the heart of the blessing.

As the Tosefta lays out the blessings associated with the sight of spectacular creations, it makes apparent a central dimension of disability experience, what Rosemarie Garland-Thomson has called, "the history of being on display, of being visually conspicuous while politically and socially erased."[61] In these rabbinic encounters, the experience of the disabled individual is entirely irrelevant to the blessing. It is the evaluative gaze of the viewer that makes the disabled body signify, that acknowledges and neutralizes the threat it poses. In theorizing the cultural power of the "freak shows" that were a hallmark of American popular culture between 1835–1940, Garland-Thomson calls attention to the way in which discourses of race, disability, animality, gender transgression, and sexual deviance are brought together in the constructed body of the freak.[62] While late antique blessing practice differs in obvious and important ways from the sideshow spectacles of late 19th and early 20th century America, Garland-Thomson's analysis of the freak show helps illuminate a central shared dynamic: the way the act of blessing fixes "the mute freak as a figure of otherness upon which the spectators could displace anxieties and uncertainties about their own identities."[63]

It is important not to overstate the cultural power of otherness. Rabbinic blessings work to preserve a place for disability and difference within the bounds of sanctified creation. Bonnie Gracer has called attention to the sharp contrast between the ideologies of infanticide in the ancient world and the rabbinic refusal to accept the killing of a disabled infant or fetus, a commitment echoed in this blessing's call to "see human variety as evidence of God's greatness."[64] Yet even as rabbinic practice affirms a place for disability, the act of blessing becomes an occasion reiterating the performativity of deviance. Through the dynamics of the blessing, disability (like racial difference) becomes an object of fascination, a site through which the normative seer perceives and ritualizes the non-normative body as dissident and deviant. Readers who reflexively associate the act of blessing with a beneficent religious act may miss the violence inherent in taking a body as an object of strangeness, fashioning the marginalized person as nothing more than an occasion for the dominant speaker to utter praise. This objectification remains familiar, in our own cultural moment. So many of us whose bodies are marked as deviant or non-conforming know the dangers of being fashioned into a symbol, a sign onto which the dominant gaze may project its fantasies and fears.

In reading rabbinic blessing practice through the lens of queer theory, I have lifted up the ways in which a queer reading illuminates the cracks and instabilities in rabbinic categories, the way it reveals the fissures in the conceptual mapping of otherness. I have underscored the way in which rhetorics of racial otherness, disability, and sexual deviance become intertwined in rabbinic discourse, bolstering and emboldening each other. But the question I close with is a more constructive one, one that rests upon José Esteban Muñoz's claim that queerness as "a mode of desiring that allows us to see and feel beyond the quagmire of the present . . . a longing that propels us onward."[65] Heeding Muñoz's call, I argue, a queer reading of this blessing must also hold out a longing for a different kind of practice, one in which the gaze does not run solely according to the dictates of dominant power. Disability studies scholars have often lauded the capacity of disability performance to refuse dehumanizing assessments of disability, to "summon the stare" in order to turn it back upon dominant culture and thereby unravel disability stigma.[66] I, too, have surfaced such impulses in rabbinic disability stories.[67] But in these texts, I find myself desiring differently. Rather than dreaming of a moment in which these deviant bodies finally force the dominant gaze to interrogate its own assumptions, I offer you a different fantasy: a longing that these bodies on the margins might find each other, that we might reach across the architectures of the pathological to fashion and feel a kinship of deviance, that we might forge together a critical solidarity between and among the despised.

WORKS CITED

Abrams, Judith Z. *Judaism and Disability: Portrayals in Ancient Texts from the Tanach through the Bavli.* Washington DC: Gallaudet University Press, 1998.

Balibar, Etienne and Immanuel Maurice Wallerstein, *Race, Nation, Class: Ambiguous Identities.* London: Verso, 1991.

Baynton, Douglas C. "Disability and the Justification of Inequality in American History," in *The New Disability History: American Perspectives*, ed. Paul Longmore and Lauri Umansky, 33–57. New York: New York University Press, 2001.

Belser, Julia Watts. "Reading Talmudic Bodies: Disability, Narrative, and the Gaze in Rabbinic Judaism," in *Disability in Judaism, Christianity and Islam: Sacred Texts, Historical Traditions and Social Analysis*, 5–27. Darla Schumm and Michael Stolzfus, eds. New York: Palgrave Macmillan, 2011.

Belser, Julia Watts and Lennart Lemhaus. "Disability in Rabbinic Judaism," in *Disability in Antiquity,* ed. Christian Laes, 434–452. London: Routledge, 2017.

Biale, Rachel. *Women and Jewish Law.* New York: Schocken Books, 1984.

Boyarin, Daniel. "A Tale of Two Synods: Nicaea, Yavneh, and Rabbinic Ecclesiology," in *Turn it Again: Jewish Medieval Studies and Literary Theory,* ed. Sheila Delany, 20–58. Eugene, OR: Wipf & Stock, 2004.

Brocco, Giorgio. "Labeling Albinism: Language and Discourse Surrounding People with Albinism in Tanzania." *Disability and Society,* Sept. 2015, 1143–1157.

Buell, Denise Kimber. *Why This New Race: Ethnic Reasoning in Early Christianity.* New York: Columbia University Press, 2008.

Butler, Judith. *Bodies That Matter: On the Discursive Limits of 'Sex.'* New York: Routledge, 1993.

———. *Gender Trouble: Feminism and the Subversion of Identity.* New York: Routledge, 1990.

Clare, Eli. *Brilliant Imperfection: Grappling with Cure.* Durham, NC: Duke University Press, 2017.

Cornell, Stephen and Douglas Hartmann. *Ethnicity and Race: Making Identities in a Changing World.* Second Edition. London: Sage Publications, 2007.

Dorff, Elliot. "Judaism and the Disabled: The Need for a Copernican Revolution," in *Healing and the Jewish Imagination: Spiritual and Practical Perspectives on Judaism and Health*, ed. William Cutter, 107–120. Woodstock, VT: Jewish Lights, 2008.

Erevelles, Nirmela. *Disability and Difference in Global Contexts: Towards a Transformative Body Politic.* New York: Macmillan, 2011.

———. "Race," in *Keywords for Disability Studies*, ed. Rachel Adams, Benjamin Reiss, and David Serlin, 145–148. New York: New York University Press, 2015.

Friedman, Shamma. *Tosefta Atikta, Pesah Rishon: Synoptic Parallels of Mishnah and Tosefta Analyzed with a Methodological Introduction* [Hebrew]. Ramat Gan: Bar Ilan University Press, 2002.

Hauptman, Judith. *Rereading the Mishnah: A New Approach to Ancient Jewish Texts.* Tübingen: Mohr Siebeck, 2005.

Garland-Thomson, Rosemarie. *Extraordinary Bodies: Figuring Physical Disability in American Culture and Literature.* New York: Columbia University Press, 1997.

———. "The Politics of Staring: Visual Rhetorics of Disability in Popular Photography," in *Disability Studies: Enabling the Humanities,* ed. Sharon L. Snyder, Brenda Jo Brueggemann, and Rosemarie Garland-Thomson, 56–75. New York: Modern Languages Association of America, 2002.

———. "Staring Back: Self-Representations of Disabled Performance Artists." *American Quarterly* 52:2 (June 2000): 334–338.

Gracer, Bonnie L. "What the Rabbis Heard: Deafness in the Mishnah." *Disability Studies Quarterly* 23:2 (Spring 2003), http://dsq-sds.org/article/view/423/593.

Gruen, Eric. *Heritage and Hellenism: The Reinvention of Jewish Tradition.* Berkeley: University of California Press, 1998.

Gurevich, Eliyahu. *Tosefta Berachot, translated into English with a Commentary by Eliyahu Gurevich.* 2010. http://www.toseftaonline.org/.

Haley, Shelley P. "Be Not Afraid of the Dark: Critical Race Theory and Classical Studies," in *Prejudice and Christian Beginnings: Investigating Race, Gender, and Ethnicity in Early Christian Studies,* ed. Laura Nasrallah and Elisabeth Schüssler Fiorenza, 27–49. Minneapolis: Fortress Press, 2009.

Hauptman, Judith. *Rereading the Mishnah: A New Approach to Ancient Jewish Texts.* Tübingen: Mohr Siebeck, 2005.

Hayes, Christine. "Legal Truth, Right Answers, Best Answers: Dworkin and the Rabbis." *Dine Israel: Studies in Halakhah and Jewish Law* 25 (2008), 73–121.

Heng, Geraldine. "The Invention of Race in the European Middle Ages I: Race Studies, Modernity, and the Middle Ages," *Literature Compass* 8/5 (2011): 315–331.

HoSang, Daniel. *Racial Propositions: Ballot Initiatives and the Making of Postwar California*. Berkeley: University of California Press, 2010.

Jagose, Anamarie. *Queer Theory: An Introduction.* New York: New York University Press, 1996.

Jastrow, Marcus. *A Dictionary of Targumim, The Talmud Babli and Yerushalmi, and Midrashic Literature.* New York: Putnam, 1903.

Kafer, Alison. *Feminist, Queer, Crip.* Bloomington: Indiana University Press, 2013.

Kessler, Gwynn. "Bodies in Motion: Preliminary Notes on Queer Theory and Rabbinic Literature," in *Mapping Gender in Ancient Religious Discourses,* ed. Todd Penner and Caroline Vander Stichele, 389–410. Leiden: Brill, 2007.

Leven, K. H. "At Times These Ancient Facts Seem to Lie Before Me Like a Patient on a Hospital Bed: Retrospective Diagnosis and Ancient Medical History," in *Magic and Rationality in Ancient Near Eastern and Graeco-Roman Medicine,* eds. H. F. J. Horstmanshoff and M. Stol, 369–386. Leiden: Brill, 2004.

Loomba, Ania. *Shakespeare, Race, and Colonialism.* New York: Oxford University Press, 2002.

Melamed, Abraham. *The Image of the Black in Jewish Culture: A History of the Other.* London: Routledge, 2003.

Muñoz, José Esteban. *Cruising Utopia: The Then and There of Queer Futurity.* New York: New York University Press, 2009.

Neis, Rachel. *The Sense of Sight in Rabbinic Culture: Jewish Ways of Seeing in Late Antiquity.* New York: Cambridge University Press, 2013.

Newman, Louis E. "'Blessed is the One Who is Good and Who Brings Forth Goodness:' A Jewish Theological Response to the Ethical Challenges of New Genetic Technologies," in *Jews and Genes: The Genetic Future in Contemporary Jewish Thought,* ed. Elliot Dorff and Laurie Zoloth, 285–309. Philadelphia: Jewish Publication Society, 2015.

Omi, Michael and Howard Winant. *Racial Formation in the United States.* New York: Routledge, 1986.

Regnault, Félix-Louis. *Hypnotisme Religion.* Paris: Schleicher Frères, 1897.

Rony, Fatimah Tobing. *The Third Eye: Race, Cinema, and Ethnographic Spectacle.* Durham, NC: Duke University Press, 1996.

Samuels, Ellen. *Fantasies of Identification: Disability, Gender, Race.* New York: New York University Press, 2014.

Satlow, Michael. *Tasting the Dish: Rabbinic Rhetorics of Sexuality.* Providence, RI: Brown University Press, 1995.

Schalk, Sami. "Critical Disability Studies as Methodology." *Lateral: Journal of the Cultural Studies Association* 6.1 (Spring 2017), http://csalateral.org/issue/6–1/forum-alt-humanities-critical-disability-studies-methodology-schalk/#fn-1748–8.

Scott, Sara. "A New Category of 'Color': Analyzing Albinism Under Title VII and the Americans with Disabilities Act." *Journal of Gender, Race and Justice* 2 (1999): 493–522.

Shoemaker, Nancy. "How Indians Got to Be Red." *The American Historical Review,* 102:3 (June 1997): 625–644.

Slifkin, Natan. *Sacred Monsters: Mysterious and Mythical Creatures of Scripture, Talmud and Midrash.* Brooklyn: Yashar Books, 2007.
Snyder, Sharon L. and David T. Mitchell. *Cultural Locations of Disability.* Chicago: University of Chicago Press, 2006.
Stein, Dina. "A Maidservant and Her Master's Voice: Discourse, Identity, and Eros in Rabbinic Texts." *Journal of the History of Sexuality* 10:3/4 (2001): 375–397.
Tuchman, Lauren. "Disability and God Talk." *Hayidion: Ravsak Journal* (Spring 2015): 18–19.
Turner, David M. *Disability in Eighteenth-Century England: Imagining Physical Impairment.* New York: Routledge, 2012.
Valler, Shulamit. *Women in Jewish Society in the Talmudic Period* [Hebrew]. Tel Aviv: Hakibbutz Hameuchad, 2000.
Whitaker, Cord J. "Race-ing the Dragon: The Middle Ages, Race, and Trippin' into the Future." *Postmedieval: A Journal of Medieval Cultural Studies* 6:1 (2015): 3–11.

NOTES

1. This chapter is written in memory of Ibrahim Farajajé, z"l.

2. Louis E. Newman, "'Blessed is the One Who is Good and Who Brings Forth Goodness:' A Jewish Theological Response to the Ethical Challenges of New Genetic Technologies," in *Jews and Genes: The Genetic Future in Contemporary Jewish Thought,* ed. Elliot Dorff and Laurie Zoloth. (Philadelphia: Jewish Publication Society, 2015), 297–8.

3. Lauren Tuchman, "Disability and God Talk," *Hayidion: Ravsak Journal* Spring 2015: 19.

4. Judith Butler, *Gender Trouble: Feminism and the Subversion of Identity.* (New York: Routledge, 1990), 112.

5. On the applicability of queer theory to rabbinic literature, see Gwynn Kessler, "Bodies in Motion: Preliminary Notes on Queer Theory and Rabbinic Literature," in *Mapping Gender in Ancient Religious Discourses,* ed. Todd Penner and Caroline Vander Stichele. (Leiden: Brill, 2007), 389–410.

6. Judith Butler, *Bodies That Matter: On the Discursive Limits of 'Sex'* (New York: Routledge, 1993), 95. Jagose argues that Butler's interpreters have often misread the concept of performativity as "free play or theatrical self-presentation" and emphasizes that Butler underscores the role that constraint and violence play in enforcing gender norms. Anamarie Jagose, *Queer Theory: An Introduction* (New York: New York University Press, 1996), 87.

7. The Tosefta, an early Palestinian rabbinic text that recounts legal material from before 200 CE, has traditionally been regarded as an explanatory companion to its canonical counterpart, the Mishnah; or as a text that contains additional material "left out" during the redaction of the Mishnah. While the Mishnah has commonly been regarded as the first text of rabbinic Judaism, several contemporary scholars have suggested that the Tosefta may actually preserve some of the oldest traditions of rabbinic legal reasoning. Shamma Friedman, *Tosefta Atikta, Pesah Rishon: Synoptic*

Parallels of Mishnah and Tosefta Analyzed with a Methodological Introduction. [Hebrew] (Ramat Gan: Bar Ilan University Press, 2002); Judith Hauptman, *Rereading the Mishnah: A New Approach to Ancient Jewish Texts.* (Tübingen: Mohr Siebeck, 2005).

8. My numbering follows the Lieberman critical edition; some translations number this passage as 6:6.

9. The Vienna manuscript also includes mention of "ha-boreq," from a Hebrew root that means "shining" or "bright." Its meaning in this context is unclear. Lieberman proposes that it is a scribal error and should be read as "ha-boheq," which he takes to indicate white scales on the skin. *Tosefta Kifshuta* 1, pg. 106.

10. The Hebrew word *qipeaḥ* is likely derived from the Greek word *kyphos*, which means "hump."

11. The printed edition, based on a manuscript that has now been lost, includes "hidraniqos," which Jastrow translates as dropsy and identifies as a corruption from the Greek "hudrópikos." Marcus Jastrow, *A Dictionary of Targumim, The Talmud Babli and Yerushalmi, and Midrashic Literature.* (New York: Putnam, 1903,) 335. The term does not appear in the Erfurt or in the Vienna manuscripts, but it does appear in the baraita that appears in Bavli Berakhot 58b. A similar term is used to reference a condition cured by Jesus on the Sabbath in Luke 14:2.

12. These categories appear in the Vienna manuscript, but not in the Erfurt manuscript or the printed edition.

13. The word is derived from the Hebrew root, "cut off." As I will discuss below, the Bavli's tradition reads it definitively as an acquired condition that affects a person's limbs. The root from which it is derived can also mean "to mutilate." Jastrow, *Dictionary,* 1351 and 1358.

14. The printed edition includes these two terms, which appear also in Bavli Berakhot 58b. They do not appear in the Vienna or the Erfurt manuscripts. The term "bohaq" appears in Leviticus 13:39, where it is identified as a white discoloration on the skin; it is distinct from the condition *tsara'at*. Jastrow suggests that this condition might be identified as albinism (143), though this does not seem to align well with the Bavli's reading of these terms as acquired conditions.

15. Ezekiel 29:10; Isaiah 20:4–6.

16. Genesis 10:6; the curse of Noah's son Canaan is articulated in Genesis 9:25–26. Though the two texts have often been conflated by exegetes who have used Genesis 9 to assert a biblical basis for black slavery, Melamed argues that the biblical text does not associate Ham or Canaan with blackness. Abraham Melamed, *The Image of the Black in Jewish Culture: A History of the Other.* (London: Routledge, 2003), 55.

17. Melamed, *Image of the Black,* 54. See Isaiah 18:1; Esther 1:1, 8:9.

18. On color prejudice in Roman society, see Shelley P. Haley, "Be Not Afraid of the Dark: Critical Race Theory and Classical Studies," in *Prejudice and Christian Beginnings: Investigating Race, Gender, and Ethnicity in Early Christian Studies,* ed. Laura Nasrallah and Elisabeth Schüssler Fiorenza. (Minneapolis: Fortress Press, 2009), 27–49.

19. Melamed, *Image of the Black,* 61–62.

20. Versions of this story appear in ySheviit 9:1 (38c), yMegilah 2:2 (73a), bRosh Hashanah 26b, and bMegillah 18a. For a discussion of the different versions, see

Shulamit Valler, *Women in Jewish Society in the Talmudic Period* [Hebrew]. (Tel Aviv: Hakibbutz Hameuchad, 2000), 152–60.

21. Dina Stein, "A Maidservant and Her Master's Voice: Discourse, Identity, and Eros in Rabbinic Texts." *Journal of the History of Sexuality* 10:3/4 (2001), 382.

22. While space does not permit me to address rabbinic discourses of animality in this chapter, the construction of "the animal" and its association with disability, race, gender, and sexual deviance is also a powerful dimension of rabbinic thought—an issue I hope to take up at length in another project.

23. Michael Omi and Howard Winant, *Racial Formation in the United States*. (New York: Routledge, 1986), 68. Daniel HoSang emphasizes that while Omi and Winant use this insight to call for "greater attention to the effects of racial power," the claim that race is not a "real" category has also been used by post-war liberal thinkers to advance the ideal of colorblindness, a framework rooted in the notion that race is an irrelevant category, that thinking about race "perpetuates the divisions our nation wishes to overcome." Daniel HoSang, *Racial Propositions: Ballot Initiatives and the Making of Postwar California,* (Berkeley: University of California Press, 2010), 243–262.

24. On the problems of retrospective diagnosis, see K. H. Leven, "At Times These Ancient Facts Seem to Lie Before Me Like a Patient on a Hospital Bed: Retrospective Diagnosis and Ancient Medical History," in *Magic and Rationality in Ancient Near Eastern and Graeco-Roman Medicine,* eds. H. F. J. Horstmanshoff and M. Stol. (Leiden: Brill, 2004), 369–386; David M. Turner, *Disability in Eighteenth-Century England: Imagining Physical Impairment* (New York: Routledge, 2012), 12.

25. While the literature on albinism remains dominated by medical and rehabilitation perspectives, scholars are increasingly conceptualizing albinism as a disability, especially in Africa. See Giorgio Brocco, "Labeling Albinism: Language and Discourse Surrounding People with Albinism in Tanzania." *Disability and Society,* Sept. 2015, 1143–1157; Sara Scott, "A New Category of 'Color': Analyzing Albinism Under Title VII and the Americans with Disabilities Act." *Journal of Gender, Race and Justice* 2:1999, 493–522.

26. Shoemaker argues that, in contrast to European and white American racial fantasies of "redskins," 18th century Native thinkers developed their own color symbolism that claimed "red" as a signifier of their own distinct cultural and political identity. Nancy Shoemaker, "How Indians Got to Be Red," *The American Historical Review,* 102:3 (June 1997), 627.

27. Eliyahu Gurevich, *Tosefta Berachot, translated into English with a Commentary by Eliyahu Gurevich.* (2010) http://www.toseftaonline.org/ Tosefta 6:6, note 4, page 341.

28. Gurevich, *Tosefta,* 341.

29. While Greek and Jewish identities are often constructed as a dichotomy in ancient and modern sources, my own thinking follows Gruen's critical insights that Jewishness and Greekness were often overlapping and interpenetrating categories. Eric Gruen, *Heritage and Hellenism: The Reinvention of Jewish Tradition.* (Berkeley: University of California Press, 1998).

30. Geraldine Heng, "The Invention of Race in the European Middle Ages I: Race Studies, Modernity, and the Middle Ages," *Literature Compass* 8/5 (2011), 265–6. I

am also indebted to Denise Kimber Buell, *Why This New Race: Ethnic Reasoning in Early Christianity*. (New York: Columbia University Press, 2008).

31. Stephen Cornell and Douglas Hartmann, *Ethnicity and Race: Making Identities in a Changing World*. Second Edition, (London: Sage Publications, 2007), 25.

32. Cornell and Hartmann, *Ethnicity and Race*, 30.

33. Heng, "Invention of Race," 265–6.

34. Cord J. Whitaker, "Race-ing the Dragon: The Middle Ages, Race, and Trippin' into the Future." *Postmedieval: A Journal of Medieval Cultural Studies* 6:1 (2015), 7.

35. Sharon L. Snyder and David T. Mitchell, *Cultural Locations of Disability*. (Chicago: University of Chicago Press, 2006), 112.

36. Douglas C. Baynton, "Disability and the Justification of Inequality in American History," in *The New Disability History: American Perspectives*, ed. Paul Longmore and Lauri Umansky (New York: New York University Press, 2001), 33.

37. Discourses of disablement are often highly paradoxical. Even as the black mind was often conceptualized as intellectually inferior, Schalk observes, "an understanding of black people's bodies as hyper strong and impervious to heat and pain also justified conditions of slave labor." Boster draws attention to the importance of attending to the material experience of disability among enslaved peoples, emphasizing the way in slaves with disabilities often recognized that manipulating prevailing assumptions about disability could prove advantageous in situations of bondage. Sami Schalk, "Critical Disability Studies as Methodology." *Lateral: Journal of the Cultural Studies Association*. 6.1 (Spring 2017), http://csalateral.org/issue/6–1/forum-alt-humanities-critical-disability-studies-methodology-schalk/#fn-1748–8. Dea H. Boster, *African American Slavery and Disability: Bodies, Property, and Power in the Antebellum South, 1800–1860*. (New York: Routledge, 2013), 4.

38. Baynton, "Disability and the Justification of Inequality," 34.

39. Nirmela Erevelles, "Race," in *Keywords for Disability Studies*, ed. Rachel Adams, Benjamin Reiss, and David Serlin. (New York: New York University Press, 2015), 147. See also Nirmela Erevelles, *Disability and Difference in Global Contexts: Towards a Transformative Body Politic*. (New York: Macmillan, 2011).

40. For a recent assessment of disability studies theory in relation to queer feminist politics, see Alison Kafer, *Feminist, Queer, Crip*. (Bloomington: Indiana University Press, 2013); on the efforts to assert and naturalize disability and racial categories, see Ellen Samuels, *Fantasies of Identification: Disability, Gender, Race*. (New York: New York University Press, 2014).

41. Etienne Balibar and Immanuel Maurice Wallerstein, *Race, Nation, Class: Ambiguous Identities*. (London: Verso, 1991), 19.

42. Fatimah Tobing Rony, *The Third Eye: Race, Cinema, and Ethnographic Spectacle*. (Durham: Duke University Press, 1996), 32, quoting Félix-Louis Regnault, *Hypnotisme Religion*. (Paris: Schleicher Frères, 1897), 161.

43. Eli Clare, *Brilliant Imperfection: Grappling with Cure*. (Durham, NC: Duke University Press, 2017), 25.

44. For a discussion of the *qipeaḥ* in terms of height, see Bavli Bekhorot 45b.

45. Pesikta Rabbati 31:4, ed. Braude, 607.

46. Though some translators have attempted to refashion this tradition apologetically, Slifkin argues that the term *parmashtak* is based on the Persian word for the penis. Natan Slifkin, *Sacred Monsters: Mysterious and Mythical Creatures of Scripture, Talmud and Midrash.* (Brooklyn: Yashar Books, 2007), 156–158.

47. yTaanit 1:6. See discussion in Melamed, *Image of the Black,* 78.

48. Ania Loomba, *Shakespeare, Race, and Colonialism.* (New York: Oxford University Press, 2002), 49–51.

49. Bavli Ketubot 60b-61a; Bavli Nedarim 20a. See Julia Watts Belser and Lennart Lehmhaus, "Disability in Rabbinic Judaism," in *Disability in Antiquity,* ed. Christian Laes. (London: Routledge, 2017), 441–442.

50. Michael Satlow, *Tasting the Dish: Rabbinic Rhetorics of Sexuality.* (Providence, RI: Brown University Press, 1995), 303–313.

51. Satlow argues that these attitudes are strikingly similar to Roman medical discourse around eugenic sexual practice. In the rabbinic corpus, these tropes are almost entirely absent in the Palestinian Talmud, but they appear in Babylonian sources, taught in the name of Palestinian sages. Satlow, *Tasting the Dish,* 312.

52. Rachel Biale, *Women and Jewish Law.* (New York: Schocken Books, 1984), 137–139.

53. Bavli Nedarim 20b. See discussion in Daniel Boyarin, "A Tale of Two Synods: Nicaea, Yavneh, and Rabbinic Ecclesiology," in *Turn It Again: Jewish Medieval Studies and Literary Theory,* ed. Sheila Delany. (Eugene, OR: Wipf & Stock, 2004), 37–43.

54. Boyarin, "A Tale of Two Synods," 43.

55. yBrakhot 9:1, 63b.

56. Judith Z. Abrams, *Judaism and Disability: Portrayals in Ancient Texts from the Tanach through the Bavli.* (Washington DC: Gallaudet University Press, 1998), 119.

57. Gurevich, *Commentary,* Tosefta Berachot 6:6, note 15, 343.

58. Christine Hayes, "Legal Truth, Right Answers, Best Answers: Dworkin and the Rabbis." *Dine Israel: Studies in Halakhah and Jewish Law* 25 (2008), 102–105.

59. Rachel Neis, *The Sense of Sight in Rabbinic Culture: Jewish Ways of Seeing in Late Antiquity.* (New York: Cambridge University Press, 2013), 9.

60. Neis, *Sense of Sight,* 8.

61. Rosemarie Garland-Thomson, "The Politics of Staring: Visual Rhetorics of Disability in Popular Photography," in *Disability Studies: Enabling the Humanities,* ed. Sharon L. Snyder, Brenda Jo Brueggemann, and Rosemarie Garland-Thomson. (New York: Modern Languages Association of America, 2002), 56.

62. Rosemarie Garland-Thomson, *Extraordinary Bodies: Figuring Physical Disability in American Culture and Literature.* (New York: Columbia University Press, 1997), 55–80.

63. Rosemarie Garland-Thomson, *Extraordinary Bodies,* 61.

64. Bonnie L. Gracer, "What the Rabbis Heard: Deafness in the Mishnah." *Disability Studies Quarterly* 23:2 (Spring 2003), http://dsq-sds.org/article/view/423/593. See also Elliot Dorff, "Judaism and the Disabled: The Need for a Copernican Revolu-

tion," in *Healing and the Jewish Imagination: Spiritual and Practical Perspectives on Judaism and Health,* ed. William Cutter. (Woodstock, VT: Jewish Lights, 2008), 109.

65. José Esteban Muñoz, *Cruising Utopia: The Then and There of Queer Futurity.* (New York: New York University Press, 2009), 1.

66. Rosemarie Garland-Thomson, "Staring Back: Self-Representations of Disabled Performance Artists." *American Quarterly* 52:2 (June 2000), 334–338.

67. Julia Watts Belser, "Reading Talmudic Bodies: Disability, Narrative, and the Gaze in Rabbinic Judaism," in *Disability in Judaism, Christianity and Islam: Sacred Texts, Historical Traditions and Social Analysis.* Darla Schumm and Michael Stolzfus, eds. (New York: Palgrave Macmillan, 2011), 5–27.

Chapter Nine

'adam Is Not Man

The Queer Body before Genesis 2:22 (and after)

Zairong Xiang

The famous "contradiction" of two creation accounts of the "Creation of Man" in the biblical Genesis has exerted a lasting puzzle for modern scholars of religion, literature, and sexuality.[1] The creation story has been commented upon by generations of theologians and biblical and literary scholars perhaps ever since these pages were written and gathered. This chapter will retake the task of analyzing the creation(s) of "man" through an agglomeration of perspectives from literary close-reading, Traditional Chinese Medicine (TCM), and "quantum queer theory." Part 1 takes Erich Auerbach's analysis of the "truth-claiming" style of the biblical narrative as a starting point to re-read, following the linear flow of the two accounts of "creation of man" in Genesis 1 and 2, while comparing different renderings in modern languages which either retain or suppress the Hebrew pun on *'adam* and Adam. Part 2 will continue to ponder the sexually entangled body of *'adam* with the help of the model of "sexless gendered body" theory that I develop, through a decolonial queer reading of TCM's body-of-orifices. Part 3 will return to the biblical text and read it together with both ancient and contemporary texts and theories (for example: Karen Barad's queer reading of quantum superposition) and suggest that the biblical text, far from proposing a sexist theology of male dominance actually retains an ambiguous space of constant queering. That "*'adam* is not man and lesbians are woman" before Genesis 2:22 therefore leaks into the "aftermath" of the primordial creation, turning the linear biblical Alpha-Omega against its heteronormative certainty. The last section will ponder on how the "queer" text *avant la lettre* might provide invaluable theory for contemporary queer rethinking of the body, ecology, and beyond.

"WOE TO THE MAN WHO DID NOT BELIEVE IT!"

Anyone who reads the biblical Genesis is likely to be struck by the two seemingly incoherent versions that recount God's creation of human beings in Genesis 1:26–27 and Genesis 2. Put in the kaleidoscope of modern languages, Genesis 1:26–27 reads as follows,

> And God said, let us make man in our image, after our likeness [. . .]; So God created man in his *own* image, in the image of God created he him; male and female created he them. (The King James Version)
>
> And God said, "Let us make a human in our image, by our likeness [. . .]"; And God Created the human in his image, in the image of God He created him, male and female He created them. (Robert Alter's Version)
>
> 天主说："让我们照我们的肖像，按我们的模样造人 [. . .]；天主于是照自己的肖像造了人，就是照天主的肖像造了人：造了一男一女。" (高思本Studium Biblicum)
>
> Puis Dieu dit: Faisons l'homme à notre image, selon notre ressemblance [. . .]; Dieu créa l'homme à son image, il le créa à l'image de Dieu, il créa l'homme et la femme. (La Bible de Jérusalem)
>
> Dijo Dios: "Hagamos al hombre a nuestra imagen y semejanza [. . .]; Y creó Dios al hombre a su imagen. A imagen de Dios lo creó. Macho y hembra los creó." (La Biblia Latinoamericana)

Translations and therefore interpretations of Genesis 1:26–27 can be grouped in two, the "man-version" and the "human-version." In the English *King James Version* and the French *La Bible de Jérusalem*, God has created "man" or "homme" according to "*our* image"; whereas in the English "Robert Alter version," the Chinese *Studium Biblicum*, and Spanish *La Biblia Latinoamericana*, God created "human," "人" (which is gender neutral) or "hombre" (which could certainly mean the gendered "male" but the version shows that "hombre" here connotes human since it uses a different word "macho" to connote male). This prose narrates God's creation of human beings according to "our"—his (her, or their)—image.

The central word here is the Hebrew *'adam*. As in many ancient Mesopotamian texts, pun is a common literary device here in the Hebrew text.[2] In the pages of the Genesis "Creation of Man" we analyze, the pun on the word *'adam* is its connotation of "human being" and the soil "*'adamah*" from which God created the human (human/humus).[3] In the commentaries, Robert Alter points out that the creation accounts (both Genesis 1:26–27 and

Genesis 2 which we will analyze later) use the word *'adam* as "a generic term for human beings [. . . that] does not automatically suggest maleness [. . .] so the traditional rendering 'man' is misleading, and an exclusively male *'adam* would make nonsense of the last clause of verse 27." [4] However, we are very familiar with the vicious use of the biblical text to justify gender inequality. In this case, the subjugation of women as purportedly supported and approved by the very creation of "man" prior to (and therefore superior to) that of "woman." However, if the misleading male gendering of *'adam* in Genesis 1:26 might be critiqued as unfounded since after all both male and female are present, the misogynist preaching on women's subordination to men seems to find its sound support in the second account of the creation: Genesis 2:19–23[5] reads,

> And LORD God fashioned from the soil each beast of the field [. . .] and brought each to the human to see what he would call it [. . .]; And the human called names to all the cattle [. . .] but for the human no sustainer beside him was found; And the LORD God cast a deep slumber on the human [. . .] and He took one of his ribs and closed over the flesh where it had been; and the LORD God built the rib He had taken from the human into a woman and He brought her to the human. And the human said: The one at last, bone of my bones and flesh of my flesh, This one shall be called Woman for from man was this one taken. (Robert Alter's Version)

> And out of the ground the LORD God formed every beast of the field [. . .] and brought [them] unto Adam to see what he would call them [. . .]; And Adam gave names to all cattle [. . .] but for Adam there was not found a help meet for him; And the LORD God caused a deep sleep to fall upon Adam [. . .] and he took one of His ribs, and closed up the flesh instead thereof; And the rib, which the LORD God had taken from man, made he a woman, and brought her unto the man; And Adam said, This [is] now bone of my bones, and flesh of my flesh: she shall be called Woman, because she was taken out of Man. (The King James Version)

At a quick glance it would seem that the biblical text is indeed androcentric if not misogynist here because no matter whether *'adam* is translated/understood as human being in general or not, we have masculine designations for it. However, the correlation between grammatical gender and bio-sociological sex is a loose one. Even if the Hebrew language genders the human (*'adam*) as grammatically masculine, it does not follow that humankind in general should start in terms of the male sex. The snake is designated as "he" in Hebrew. However the grammatically masculine "snake" has not impeded scholarly and artistic receptions from associating it with the feminine[6] because of the alleged "complicity" between the snake and the woman.[7] In

Hebrew, "the nouns that have biological gender features are often assigned a grammatical gender that does not reflect true gender distinctions."[8]

The male gendering of God's first creature has been further fixed in an instant when God in the King James Version (KJV) and in the Latin Vulgate, suddenly calls upon the human with the male name "Adam" in Genesis 2:19. God brings the animals for Adam to name them, and from *"his"* rib God creates the woman. Originally used as a pun in the Hebrew text, *'adam* the generic human echoing *'adamah*-the earth[9] from which *'adam* is made, the word is here rendered in translation as a proper name "Adam," gendered male.

Although it might sound perfectly "normal" for the male Adam to be created first, it does not make sense: to have a "male" while a "female" does not yet exist. When we put the two biblical narratives of the creation of *'adam* within its narratology, to translate and therefore interpret—moving from the linguistic to the social—*'adam* as Adam (a male name, suggesting "his" maleness) can be hardly a viable option. In the first creation account, we read "Let us make a human in our image, by our likeness [. . .] And God created the human in his image, in the image of God He created him, male and female He created him" (Genesis 1:26–27). Alter notes that *'adam* consists of a definite article and refers to humankind instead of an individual man named Adam. Consequently the pronoun "him" referring to *'adam* the human is grammatically masculine but it does not automatically connote maleness. Otherwise, Genesis 1:27 in which male *and* female are created would not have made sense.

However, Alter contends that the feminist suggestion that "*'adam* is to be imagined as sexually undifferentiated until the fashioning of woman" would lead to "certain dizzying paradoxes in following the story."[10] I understand this "dizzying paradox" as either the impossibility of envisaging *'adam* beyond male/female or the absurdity of having first an undifferentiated *'adam* and then a masculinized one later. If one brings in an important feature of biblical narratology into consideration, this impossibility or absurdity might neither be dizzying nor paradoxical. Eric Auerbach in his seminal work *Mimesis: The Representation of Reality in Western Literature* by comparing the Homeric and the biblical narrative strategies argues that "the Biblical narrator was obliged to write exactly what his belief in the truth of the tradition [. . .] demanded of him [. . . and what] he produced, [. . .] was oriented toward truth."[11] That is to say, this claim to absolute truth does not need or cannot rely on narrated "reality" but is achieved by what Auerbach calls a tyrannical gesture, "Woe to the man who did not believe it!"[12]

In light of this, let's reassess what Alter deems to be dizzyingly paradoxical. Ronald Hendel suggests that "sex and gender is not an emphasis of the identity of Adam [and . . .] Adam before the creation of woman is in a sense 'not yet' specifically male."[13] If we strictly follow the sequence of the biblical narration with its "truth claiming" logic in mind, we are not told yet about the creation of woman. Only until the narrator informs us in Genesis 2:23 do we know the existence of the woman. It is only after the creation of a *woman* that *'adam* can become known as Adam, leaves "his" generality behind, and becomes a man. The ancient text skillfully plays against our "common sense" and the habit of assuming male priority and sexual dichotomy. With the help of the biblical text, we *can* convey the idea of a sexless (grammatically) gendered *'adam* without much ado.

'adam AS YELLOW EMPEROR'S BODY-OF-ORIFICES

Before Genesis 2:22, *'adam* is not man. In fact, if we strictly follow the biblical "truth claiming," we are not supposed to notice the different reproductive genitals before the couple eats the forbidden fruit and discovers that they are naked so that "they sewed fig leaves and made themselves loincloths" (Genesis 3:7). Before this moment in which our attention is drawn to the genital parts of their bodies that mark their sexual differences, not only *'adam* but also Adam and the woman are all *not yet* sexed or gendered. As we will see in the next part, the biblical text, largely outside of what we could call modern/colonial heteronormativity, needs some extra ink and institution to ensure the upcoming sexual differentiation and gender assignment.

Now let us dwell a little bit at the originary moment of *'adam*'s creation. Before they put on the loincloth and for the first time alienate themselves from the flora and fauna of the Eden garden, they were an organic part of "nature": from *adamah*—soil *'adam* the human is made. Walking in the blissful paradise of primordial innocence, *'adam* is naked, Adam and the woman are all naked or in fact, "nude." John Berger makes a famous distinction in Western art between naked and nudity that could be useful here. "To be nude is to be seen naked by others and yet not recognized for oneself. A naked body has to be seen as an object in order to become a nude."[14] Not unlike the ideologically saturated concept of "nudity," which is an intricate technology of gender in Western visual representation, *'adam*'s nakedness (at least before the "fall" in Genesis 3) can in fact only be "nude" in the Bergerian sense of the term: their status as an object and their unawareness of their nakedness.

The nude *'adam* is not separate from the flora and fauna of the world recently created. The boundary between them is impossible. After all, they are all fashioned from the earth. In this sense, *'adam* is permeable, part of the landscape. In order to better understand this bodily porosity, we will visit an old Chinese doctor Qibo. Chapter 4 "Jinkui zhenyan" [金匮真言, or the Truth from Golden Closet] of the medical classic *Huangdi Neijing* [黄帝内经, or *Yellow Emperor's Inner Canon*] records Qibo the erudite doctor's detailed answer to the question Yellow Emperor Huandi asks: "The five depots correspond to the four seasons. Do all of them have [something specific] that they collect and receive?" [五脏应四时，各有收益乎？].[15] Qibo answers this question about the connectivity between human body (five depots) and the four seasons using a porous model that could be called the body-of-orifices. Equally an ancient text written long before the modern subject takes up the phallic and colonial project of subjugation of the world, *Huangdi Neijing* has a theory of the body that is based on porosity.[16] Let us leave the Hebrew garden for a moment and look at this Chinese text.

Qibo constructs a theory of body that is based on the connectivity between the inner organs or "five depots"(五脏) and the nine bodily orifices (窍).[17] One particular organ draws my attention, the "yin-orifice":

> The North; black color. Having entered it communicates with the kidneys. It opens an orifice in the two yin [sites] [18]
>
> 北方黑色，入通于肾，开窍于二阴。[19]

In this account, the body-of-orifices is sexless. The "two yin orifices" are the anus and genitalia *regardless of* sexual difference. That is to say, Qibo's theory of the body sees both the penis and the vagina as the same (yin) orifice. Retheorizing the body against its heterosexualist understanding is an important task in contemporary queer theory.[20] "Conventional heterosexual nonreciprocity depends upon a 'hegemonic bodily imago of masculinity' that conforms with the man's status as sovereign ego and with an understanding of the male body as 'phallic and impenetrable.'"[21] This phallic and impenetrable bodily imago is however betrayed by the male body itself, especially with the anus. Paul B. Preciado succinctly shows the power of the anus in this regards through a pun in Spanish. "Por el ano, el sistema tradicional de la representación sexo/género se caga" (Through the anus, the traditional system of sex/gender representation shits itself or is shat upon).[22]

While queer theorists working almost exclusively within a limited reference of Western traditions busily take (male) anality to task and theorize the rectum or the "lesbian phallus,"[23] Yellow Emperor's porous body might be useful for further radicalizing contemporary queer theory which has left the penis' phallic status unquestioned. The penis just like vagina and anus, as well as the eyes, the nostrils and the mouth, is but a penetrable hole. One only needs to look at it closely to agree, with the urethra and sexual practices such as "sounding" to concur with the porous rendering of the penis.

The penis and the vagina are both called "yin qiao (阴窍)" or the orifices with yin attributes. However, they are not the two orifices where the kidney connects to, which, according to the "Truth from the Golden Closet" are the frontal yin orifice (前阴) and the rear yin orifice (后阴), namely the genital and the anus. Penis and vagina in their capacity as the "frontal orifices" resume yin attributes and is seen as one and the same hole. This does not mean that Yellow Emperor's body-of-orifices is sexless. It is not even androgynous as Charlotte Furth has argued.[24] The sexual differentiation familiar to the modern heteronormative understanding of the body is available. For example, penis is also referred to as the "yang tool" (阳具) and vagina the "yin path" (阴道). When put side by side, the frontal orifices get their respective yin-yang attributes.[25] In another words, the penis and vagina stops differentiation and becomes superimposed as one: the frontal orifice (前阴), when the anus, that is, the rear orifice (后阴) is brought up. The frontal and rear orifices together in their capacity as the porous connectivity of the kidney become simply "two yin orifices" (二阴).

This is not the end of the story, but only a start. According to the *Inner Canon*, there are simultaneously several cosmic pairs at different levels of the universe. What we have looked at with the bodily orifices is the level of "sexual body" in which the menses assume the yin attribute as yin-blood while the semen assumes the yang attributes as the yang-essence. Semen and menses of the "sexual body" at a different level become one thing known as the "yin-blood" to be differentiated from and combined with the "yang-essence" that has no bodily bearing.[26] This level is further "subsumed" as one "yin-essence," differentiated from and combined with a cosmic "yang-qi" as the generative power of the earth. If we take yin-yang attributes as gender assignments, the *Inner Canon* consists of a system of relativism of gender differentiation (rather than differences).

These pairings that change yin-yang status at different levels in relation to each other in a certain situatedness, always subsume the yin-yang attributes into a status of "wave-particle" superposition (to borrow a concept from quantum physics).[27] I have called this simultaneous differentiation and combination of yin-yang attributes in the body elsewhere a "transdualism of either . . . and," keeping the differentiating propensities ("either") while stressing their porosity, inter-becoming and "intra-action" ("and").[28]

Back to the penis. This "cavity" inside, at the core of the almost undeniable "male organ" so often conflated with the impenetrable phallus enables us to suspend the "hard facts" of sexual differences that heteronormative biological essentialism finds so useful, into the indeterminacy of a cosmic body that is either yin *and* yang, either feminine *and* masculine. This at once abstract and concrete body is permeable. Its inner organs are connected with the outer orifices but also between with the four seasons (as well as the five cardinal directions, the five elements, the five colors, the five tastes, the five musical notes). They *is* free in the cosmic garden of Eden.

Now let us go back to the suspended moment of the "dizzying paradox" that *'adam* should be seen as sexually undifferentiated before the creation of woman. If we regard *'adam* as Yellow Emperor's body-of-orifices that is either "sexless" *and* "gendered" (yin-yang attributed) on different generative levels at the scale of the universe, we might understand *'adam*, fashioned from *adamah* the earth, as a body of superposition that is not male or female or androgynous (yet). This superposition only gets "resolved" with the creation of the woman from *'adam*'s rib. They *is* an electron of indeterminacy. These grammatical discordances between the subject and the predicate are hardly scandalous. The very first sentence of Genesis already has an intensively queer poetic juxtaposition and disruption on play. *bara Elohim* uses the third person singular "bara" for a plural noun "Elohim." Catherine Keller (one of the other authors in this volume) coins the term "pluri-singular" for *elohim.*[28] And if God "himself" is at once plural and singular, the one(s) created in "his" image would follow this dizzying jump between the plural and the singular, captured precisely by the first account of the creation: "And God Created [*bara Elohim*] the human in his image, in the image of God He created him, male and female He created them" (Genesis 1:26–27). "God 'himself' is unsure whether he is plural or singular [. . . and the] blurred and slipping self-definition suggests that things [. . .] might in fact be as inherently indeterminable as the identity that creates it."[29]

'adam IS NOT MAN

> And the rib, which the LORD God had taken from man, made he a woman, and brought her unto the man. And Adam said, This is now bone of my bones, and flesh of my flesh: she shall be called Woman, because she was taken out of Man.
>
> Therefore shall a man leave his father and his mother, and shall cleave unto his wife: and they shall be one flesh. And they were both naked, the man and his wife, and were not ashamed. (Genesis 22–25, KJV)

'adam recently wakes up with one rib taken which has become a woman. He claims her to be bone of his bones, flesh of his flesh, and in a typical androcentric manner: "she shall be called Woman, because she was taken out of man." Immediately afterwards however, as if unsure which "gender

ideology" to promote, the Genesis text reads: "Therefore shall a man leave his father and his mother, and shall cleave unto his wife."

After the creation of the woman, the biblical text deems it necessary to introduce the heterosexual arrangement. This arrangement however, far from laden with patriarchal cliché, requires man rather than woman to leave his parents and "cleave unto his wife." That is to say, in two neighboring verses at this primordial moment, there is no suggestion of the "traffic in women."[30] The moving direction of the man towards the woman troubles the androcentric certainty as well as the facile condemnation of biblical (or at least Genesis') misogyny. It is coherent that the snake later should speak first and directly to Eve (the one Adam has to cling to) but not to Adam, in order to seduce them into trespassing God's prohibition. Eve, the *'ezer* given to *'adam* by God in Genesis 2 is a "sustainer besides him"[31] rather than merely "an help meet for him" (KJV) which might suggest an inferior position in the hierarchy.

And "they become one flesh" (Genesis 2:24). Alter comments, "characteristically, the narrative immediately unsettles the neatness of the etiological certainty for the first couple are two, not one flesh" (1995:10n25), due to the next verse: "And the two of them were naked, the human and his woman, and they were not ashamed" (Genesis 2:25). As we have seen in the previous two parts, the question of two and one is rather complex in the biblical creation of *'adam*. It is misleading to understand the two as "two, not one flesh" as Alter suggests. Genesis at this point at least, is rather uncertain about the couple. What we have is the "bone of bones, flesh of flesh" one-ness. Is this couple consisting of either one *and* two "subjects," very much in the same manner of "bara Elohim," a pluri-singular intra-action?

Karen Barad (mentioned in many of the chapters of this volume) detects certain queerness of the "quantum leap" of the electron as "a dis/continuous movement, and not just any discontinuous movement, but a particularly queer kind that troubles the very dichotomy between discontinuity and continuity." This queerness according to her "troubles the very notion of *dicho-tomy*—the cutting into two—itself (including the notion of 'itself'!)."[32] It is hardly meaningful to suggest that *'adam* before Genesis 2:22 (or the *Inner Canon* for that matter) has intuited a kind of quantum theory, long before the spooky nature of electrons has been theorized. However, the "quantic *'adam*" urges us to thoroughly revise the male/female dichotomy, a strong "gender ideology" through which we have been told to reading the Genesis text and human subjects. That is, before rushing to queer an allegedly heteronormative text that has sponsored queerphobia, we could learn from the juxtaposition of Genesis 2:24 and 2:25 as a warning against the modern/colonial habit of dichotomization, which is at the same time hierarchization. Already embedded in the text is perhaps anachronically a queer strategy at odds with both "nihilistic dissolution of meaning on the one hand and unification by self-appointed orthodoxy on the other."[33]

Then Adam and the woman ate the forbidden fruit. They put on the loincloth to cover but explicitly the genitals after having their eyes open. It awakes their awareness of their own bodies, turning the nude into the naked. "The woman" eventually gets her name, Eve. We notice that it is only after the "fall" and God's curse on Adam, do we first encounter her with a proper name. Before Genesis 2:23 *'adam* is not a man and lesbians *are* women[34] paraphrasing Monique Wittig.[35] The authors of the Hebrew Genesis seem to agree with Wittig if not with her "critique of heteronormativity," for they are quite aware that neither "man" nor "woman" would make sense outside heterosexuality. Wittig provocatively states that "lesbians are not woman." What she proposes is a critique of what is later known as heteronormativity, or as she calls it, the "straight mind." The queer recategorization and decategorization in Wittig is similarly pondered by Eve Sedgwick when she enumerates the possible "identities" that are not based on a boring, simple, and at times murderous woman/man dichotomy: "pushy femmes, radical faeries, fantasists, [. . .] ladies in tuxedoes [. . .] leatherfolks [. . .] lesbian-identified men or lesbians who sleep with men."[36] In light of this and to push the queer/quantic suspicion of category or categorial certainty to its extremity, "lesbians are not woman" and "lesbians are woman," achieve a similar effect. They both disrupt the certainty and expose the fallacy of the category "woman" and also "lesbian" that have been largely understood as culturally universal, historically transversal, and biologically self-evident.

SOME THOUGHTS AFTER GENESIS 2:22

Contemporary critical theories, especially queer theorization have taken the task of seeking not only to recognize the diversity and multiplicity of life, but most importantly to promote pluriversal onto-epistemologies that depart from and are able to theorize the ultimately uncategorizable, queer, and largely repressed nature of nature and society. It is my intention to move to a different direction: bringing these pluralist luxuriances back to originary moments imbued with dizzying paradoxes.[37] The biblical Genesis delights in using puns, the extraordinary ability of language, to capture and also enact the multiplicating creational energy: *'adam, Adam, adamah, 'adham.* "Be fruitful and multiply!" (Genesis 1:29). The *Inner Canon* resolves the ultimately unrepresentable multitude by emptying out the ontological certainty of proper name. Yellow Emperor's cosmic body insists on yin-yang indeterminacy by collapsing all possible variants into an all-encompassing porosity of the universe of which the materio-discursive body-of-orifices partakes. Quantum theory works with this originary paradox in terms of entanglement. Karen Barad claims via Bohr, "our knowledge-making practices are material enactments that contribute to, and are part of, the phenomena we describe."[38] To address this entanglement is part of the political and ethical endeavors of contemporary theorizations. Trible notes, after being expelled from the Garden of Eden, "the conclusion of the story does not specify the sexes in flight. Instead the narrator resumes use of the generic and androgynous term *'adham,* with which the story began, and thereby completes an overall ring composition."[39] If *'adam* is not man before Genesis 2:22, which potentially renders *their* body "queer," the queerness further seeps through the temporal closet of linear *straight*-forwardness, haunting the thereafter of the subject formation, the individually sexed humans now concretely named Adam and Eve.

WORKS CITED

Alter, Robert. *Genesis: Translation and Commentary*. London: W. W. Norton & Company Inc., 1996.

Auerbach, Eric. *Mimesis: The Representation of Reality in Western Literature*. Princeton: Princeton University Press, 2003.

Barad, Karen. *Meeting the Universe Halfway: Quantum Physics and the Entanglement of Matter and Meaning*. Durham: Duke University Press, 2007.

———. "Quantum Entanglements and Hauntological Relations of Inheritance: Dis/continuities, SpaceTime Enfoldings, and Justice-to-Come." *Derrida Today* 3, no. 2 (2010): 240–68.

Berger, John. *Ways of Seeing*. London: Penguin, 2008.

Bersani, Leo. *Is the Rectum a Grave?: And Other Essays*. Chicago: University of Chicago Press, 2010.

La Biblia Latinoamericana. http://www.sanpablo.es/biblia-latinoamericana. Last Accessed December 15, 2017.

Biblia Sacra Vulgata. http://biblehub.com/vul/genesis/1.htm Last Accessed December 15, 2017.

Butler, Judith. *Gender Trouble: Feminism and the Subversion of Identity*. London: Routledge, 1999.

Coffin, Edna Amir, and Shmuel Bolozky. *A Reference Grammar of Modern Hebrew*. Cambridge: Cambridge University Press, 2005.

Feng, Xiang. 创世纪：传说与译注 *(Genesis: Legends and Commentaries)*. Nanjing: Jiangsu People's Publishing House, 2004.

Fewell, Danna Nolan, and David M. Gunn. *Gender, Power & Promise: The Subject of the Bible's First Story*. Nashville: Abingdon, 1993.

Freedman, David Noel, ed. *Eerdmans Dictionary of the Bible*. Grand Rapids, MI: Wm. B. Eerdmans Publishing Co., 2000.

Furth, Charlotte. *A Flourishing Yin: Gender in China's Medical History, 960–1665*. Berkeley: University of California Press, 1999.

Keller, Catherine. *Face of the Deep: A Theology of Becoming*. London: Routledge, 2003.

Keller, Catherine, and Laurel C. Schneider, eds. *Polydoxy: Theology of Multiplicity and Relation*. New York: Routledge, 2011.

The King James Version. https://www.biblegateway.com. Last accessed December 15, 2017.

Le Cerf, ed. *La Bible de Jérusalem*. Paris: Les éditions du cerf., 2007.

Lugones, María. "Heterosexualism and the Colonial/Modern Gender System." *Hypatia* 22, no. 1 (2007): 186–209.

Miller, James. *China's Green Religion: Daoism and the Quest for a Sustainable Future*. New York: Columbia University Press, 2017.

Noegel, Scott B., ed. *Puns and Pundits: Word Play in the Hebrew Bible and Ancient Near Eastern Literature*. Bethesda: CDL Press, 2000.

Preciado, Beatriz. *Manifesto Contra-Sexual*. Madrid: Editorial Opera Prima, 2002.

Rubin, Gayle. "The Traffic in Women: Notes on the 'Political Economy' of Sex." In *Toward an Anthropology of Women*, 157–210. New York: Monthly Review Press, 1975.

Sedgwick, Eve Kosofsky. *Tendencies*. London: Routledge, 1994.

高思本 Studium Biblicum. http://www.chinacatholic.net/txt/bible/. Last Accessed December 15, 2017.

Thomas, Calvin. "Reenfleshing the Bright Boys; Or, How Male Bodies Matter to Feminist Theory." In *Masculinity Studies and Feminist Theory: New Directions*. Edited by Judith Kegan Gardiner, 31–59. New York: Columbia University Press, 2002.

Trible, Phyllis. "Depatriarchalizing in Biblical Interpretation." *Journal of American Academy of Religion* 41, no. 1 (1973): 30–48.

Unschuld, Paul U., Hermann Tessenow, and Jinsheng Zheng, trans. *Huang Di Nei Jing Su Wen: An Annotated Translation of Huang Di's Inner Classic—Basic Questions*. Berkeley: University of California Press, 2011.

Wittig, Monique. "One Is Not Born a Woman (1981)." In *The Straight Mind and Other Essays*, 9–20. Boston: Beacon Press, 1992.

Xiang, Zairong. *Queer Ancient Ways: A Decolonial Exploration*. New York: Punctum Books, Forthcoming.

NOTES

1. This research derives from a research project of the author's during his tenure as a research fellow at the ICI Berlin Institute for Cultural Inquiry.

2. See: Scott B. Noegel, ed., *Puns and Pundits: Word Play in the Hebrew Bible and Ancient Near Eastern Literature* (Bethesda: CDL Press, 2000).

3. Robert Alter, *Genesis: Translation and Commentary* (London: W. W. Norton & Company Inc., 1996), 8n7.

4. Ibid., 5.

5. Given the length of the passage, I will only quote the two English versions.

6. See for example Michelangelo’s *Temptation* and the representation of Genesis on the façade of Notre Dame de Paris.

7. Xiang Feng points out that we can be sure, from the narrative of Genesis 3:4, that Adam is present when the snake talks to the woman. Otherwise, it would not have been possible that “she took of its fruit and ate, and she also gave to her man, and he ate” (Genesis 3:6). The age old condemnation of the first woman’s sole responsibility for “seducing” the first man to eat the forbidden fruit. Feng, 创世纪：传说与译注 *(Genesis: Legends and Commentaries)*. See also: Phyllis Trible, “Depatriarchalizing in Biblical Interpretation,” *Journal of American Academy of Religion* 41, no. 1 (1973),

8. Edna Amir Coffin and Shmuel Bolozky, *A Reference Grammar of Modern Hebrew* (Cambridge. Cambridge: Cambridge University Press, 2005), 133.

9. Alter, *Genesis: Translation and Commentary*, 8n7.

10. Ibid., 5.

11. Eric Auerbach, *Mimesis: The Representation of Reality in Western Literature* (Princeton: Princeton University Press, 2003), 14.

12. Ibid.

13. David Noel Freedman, *Eerdmans Dictionary of the Bible* (Wm. B. Eerdmans Publishing Co., 2000), 18–19

14. John Berger, *Ways of Seeing* (London: Penguin, 2008), 54.

15. English translation quoted from: Paul U. Unschuld, Hermann Tessenow, and Jinsheng Zheng, trans. *Huang Di Nei Jing Su Wen: An Annotated Translation of Huang Di’s Inner Classic—Basic Questions*. Berkeley: University of California Press, 2011), 91.

16. TCM is not the only Chinese tradition that has a theory of the porous body. Daoism for example has elaborated theory and practice regarding the porosity of the body. See: James Miller, *China’s Green Religion: Daoism and the Quest for a Sustainable Future* (New York: Columbia University Press, 2017), 66–88.

17. East—Green—Liver—Eyes; South—Red—Heart—Ears; Centre—Yellow—Spleen—Mouth; West—White—Lung—Noses; North—Black—Kidney—Two Yin Orifices: genital and anus. See also: Unschuld, Tessenow, and Zheng, *Huang Di Nei Jing Su Wen*, 91–93.

18. English translation quoted from: Ibid., 93.

19. In chapter five, it also connects to the ears.

20. “Heterosexualism” is a concept proposed by Maria Lugones to address the coloniality of heteronormativity. I follow her to use this term to remind the reader of the modernity/coloniality critique in the analysis of heteronormativity. See: Lugones, “Heterosexualism and the Colonial/Modern Gender System.”

21. Calvin Thomas, “Reenfleshing the Bright Boys; Or, How Male Bodies Matter to Feminist Theory,” in *Masculinity Studies and Feminist Theory: New*

Directions, ed. Judith Kegan Gardiner (New York: Columbia University Press, 2002), 31–59.

22. Preciado, *Manifesto Contra-Sexual*. English translation mine.

23. Leo Bersani, *Is the Rectum a Grave?: And Other Essays* (Chicago: University of Chicago Press, 2010); Judith Butler, *Gender Trouble: Feminism and the Subversion of Identity* (London: Routledge, 1999).

24. See: Charlotte Furth, *A Flourishing Yin: Gender in China's Medical History, 960–1665* (Berkeley: University of California Press, 1999).

25. That is to say, borrowing from the vocabulary of quantum physics: "the point is that measurement resolves the indeterminacy," they are always together in a status of superposition, only measurement or when they are attributed, resolves this originary status of indeterminacy. See: Barad, "Quantum Entanglements and Hauntological Relations of Inheritance: Dis/continuities, SpaceTime Enfoldings, and Justice-to-Come," 280.

26. See: Furth, *A Flourishing Yin: Gender in China's Medical History, 960–1665.*

27. See: Barad, *Meeting the Universe Halfway: Quantum Physics and the Entanglement of Matter and Meaning*; and Barad, "Quantum Entanglements and Hauntological Relations of Inheritance: Dis/continuities, SpaceTime Enfoldings, and Justice-to-Come."

28. Catherine Keller, *Face of the Deep: A Theology of Becoming* (London: Routledge, 2003), 172.

29. Danna Nolan Fewell and David M. Gunn, *Gender, Power & Promise: The Subject of the Bible's First Story* (Nashville: Abingdon, 1993), 23.

30. Gayle Rubin, "The Traffic in Women: Notes on the 'Political Economy' of Sex," in *Toward an Anthropology of Women* (New York: Monthly Review Press, 1975).

31. *'ezer*, which is usually translated as "helper" as an indication of Eve's inferiority to Adam, is also silently contested by the original text. According to Phyllis Trible, "*'ezer* is a relational term; it designates a beneficial relationship; [. . . and b]y itself the word does not [. . .] imply inferiority" Trible, "Depatriarchalizing in Biblical Interpretation," 1973.

32. Barad, "Quantum Entanglements and Hauntological Relations of Inheritance: Dis/continuities, SpaceTime Enfoldings, and Justice-to-Come," 246.

33. Keller and Schneider, *Polydoxy: Theology of Multiplicity and Relation*, 3.

34. Wittig's provocative claim that "lesbian is not woman" intends to question the "compulsory heterosexuality" underpinning the gender dichotomization of "man" and "woman." Before Genesis 2:23, that is to say, before the creation of "woman," neither "man" nor "woman" would have a meaning. In this particular moment, aligned with Wittig's critique, lesbians can be "woman" (a concept that has not yet been heteronormativized).

35. Monique Wittig, "One Is Not Born a Woman (1981)," in *The Straight Mind and Other Essays* (Boston: Beacon Press, 1992), 9–20.

36. Eve Kosofsky Sedgwick, *Tendencies* (London: Routledge, 1994), 8.

37. I have discussed in detail elsewhere the "queerness" of originary moments in different cultural texts by rereading two different creation myths in: *Queer Ancient Ways: A Decolonial Exploration*. (New York: Punctum Books, forthcoming).

38. Barad, *Meeting the Universe Halfway: Quantum Physics and the Entanglement of Matter and Meaning*, 247.

39. Trible, "Depatriarchalizing in Biblical Interpretation," 1973, 42.

Chapter Ten

Gender and Indeterminacy in Jewish Mystical Imagery

Fern Feldman

Indeterminacy, a concept developed by physicist Niels Bohr, and further elaborated by philosopher Karen Barad, tells us that there are no fixed, preexistent separate entities, but rather entities arise out of the phenomena of which they are part. If we accept, as recent physics research has shown, that quantum ontology functions at all scales,[1] we can expect gender (and other social variables) to be susceptible to indeterminacy as well. Gendered beings arise out of a relational matrix involving a multitude of changing forces (social, cultural, economic, ecological, physical, and more), and this process is visibly at play throughout Jewish sacred text and mystical traditions.

Transgender, genderqueer, and intersex people have raised awareness that gender and sex are not binary systems. Inspired by feminist and queer theory, we no longer need to associate women with body and emotions, and men with mind and logic, nor overlay other binaries onto gender identities. Not only gender, but binaries of all sorts are reified in much popular spiritual imagery—the transcendent divine is associated with the male aspect of divinity, up above the female/immanent aspect. The male is imagined as giver, the female as receiver, in an intra-divine heterosexual coupling. But Kabbalah (Jewish mysticism) contains multifold diverse images for the divine interplay that go beyond the stereotypical missionary position.

This chapter reads Hebrew Scriptures and Jewish mystical texts with Judith Butler's theory of gender performativity, Barad's theory of agential realism, and Bohr's concept of indeterminacy, exploring queer and genderqueer images and paradigms within the Jewish textual tradition, and examining how aspects of the divine flow shift gender, gender role, and orientation in a rich interplay to create life, nourish the world, and bring redemption.

APPROACHES TO COMPLICATING THE GENDER BINARY

Spatial Reversal

A variety of approaches within Jewish mysticism complicate the stereotypical binaries of gendered imagery. One approach is exemplified in much Chasidic writing. The 18th century Chasidic book *Toldot Ya'acov Yosef*, states "Thus man becomes sanctified to be a chariot for the Shekhina, in body and soul, as (he was) at the Revelation at Mount Sinai . . . This is the true wedding ceremony, and understand it."[2] Here, there is a spatial reversal, and the male is on the bottom. The pair that is focused on is the immanent divine and the (presumably male) person engaged in prayer or study. The male religious practitioner becomes the passive receiver, and a female aspect of the divine, the active giver of influx. This is a powerful archetype that has influenced, and been influenced by, how Jews see Jewish men.[3] What it means to be male, then, is not just one half of a duality, where male=top, but rather there is an acknowledgment that individuals have more valences to who they are, and how they participate in the dance of giving and receiving.

The Jewish image of man as receiver is rooted in much earlier tradition. For example, the medieval Midrashic[4] collection *Exodus Rabbah* (45:3) teaches:

> You find sometimes, "And Adonai spoke to Moses," and, "And Adonai said to Moses"; so also you find, "And Moses said to Adonai," and also "And Moses spoke to Adonai." It can be compared to a cave situated by the seashore into which the sea once penetrated, and having filled it, never departed, but was always flowing in and out of it.

Here, Moses, the Torah's greatest teacher and prophet, is being described as a cave, a vessel, filled with divine presence, male human imagined as the passive receiver penetrated and filled with divine flow.

Disconnecting "Male" from Men and "Female" from Women

A second approach to altering the gender binary involves pointing out that *actual* men are not just "male" and *actual* women not just "female." This sort of view is expressed by contemporary author Sarah Schneider, in her book, *Kabbalistic Writings on the Nature of Masculine and Feminine*. She writes "Everything is feminine in relation to its above and masculine in relation to its below. There is no person who is absolutely and exclusively male or female, for there is no escaping this network of relations with its shifting hierarchies and gender roles that vary from one context to the next." [5]

This perspective addresses the problem of societal inequalities that place men on top, and women on the bottom, by pointing out that who is in the "male" position and who the "female" is not dependent on the plumbing of the individual, but rather shifts and varies. Perspectives like Schneider's open up our understanding of gender in many ways, as well as being helpful for some people who experience themselves as gender fluid. But there is still a perpetuation of stereotypes of what "male" and "female" mean. It does nothing to trouble the problematic equations of male equals top/active/giver and female equals bottom/passive/receiver, nor the essentialist and binary approach to gender itself that is the norm, nor does it question the assumption that all interactions involve a male and a female participant.

Inseparability of Sex and Gender

A third approach is to claim that a binary approach is necessary, due to the biological sex differences between men and women.[6] Often, there is a distinction made between sex and gender, where gender is taken to be cultural, and sex biological. But Judith Butler argues that sex is actually inseparable from gender. She writes:

> Sexual difference . . . is never simply a function of material differences, which are not in some way both marked and formed by discursive practices. Further, to claim that sexual differences are indissociable from discursive demarcations is not the same as claiming that discourse causes sexual difference. The category of "sex" is, from the start, normative; it is what Foucault has called a "regulatory ideal." In this sense, then, "sex" not only functions as a norm, but is part of a regulatory practice that produces the bodies it governs, that is, whose regulatory force is made clear as a kind of productive power, the power to produce—demarcate, circulate, differentiate—the bodies it controls. Thus, "sex" is a regulatory ideal whose materialization is compelled, and this materialization takes place (or fails to take place) through certain highly regulated practices. In other words, "sex" is an ideal construct which is forcibly materialized through time. It is not a simple fact or static condition of a body, but a process whereby regulatory norms materialize "sex" and achieve this materialization through a forcible reiteration of those norms.[7]

Butler teaches that sex cannot be taken as a given, as some essential inherent determined and determining binary. Rather, it is "part of the regulatory practice that produces the bodies it governs."[8] This chapter will question the regulatory practices that compel conformity as it explores the multiple valences of gendered imagery in Kabbalah and Jewish sacred text that conceptualize divine creative flow.

The Lovers as Receivers, Together

Rabbi Zalman Schachter-Shalomi, z"l in his essay Sameach T'samach,[9] on sacred sexuality, presents an interesting approach. He writes (in Hebrew, my translation): "'When a female surrounds a male'[10] there is not any recognizable separation between the giver of the flow and the receiver, because each one says that s/he is receiving the joy of the bestowal of love from the other." Shachter-Shalomi then goes on to cite the Song of Songs: "O! That you were like a brother to me, nursing at my mother's breasts! If I should find you outside, I would kiss you; and none would despise me" (8:1). He comments, "nursing—*together with me*—the breasts of my mother." The image he develops here is a creative approach. He attributes agency to the female—instead of being a passive recipient of penetration, the female is said to "surround" the male, in an active way. (Here, he cites Jeremiah, so this image of female agency is rooted in the Tanakh, the Hebrew Bible, itself). In addition, instead of one partner being above and sending flow to the other, both partners are below, receiving the flow together from the supernal source, the divine Mother, *Binah Ila'a*, the supernal *Binah* (uppermost matriarch of the sefirotic Tree of Life). We will return to this image later.[11]

Queer Intra-Divine Coupling

Another approach to queering the gendered imagery of Kabbalah shifts the focus away from solely heterosexual relations. This was explored at least as early as 13th century Spain, by Joseph Gikatilla, in his book, *Sha'arei Orah, Gates of Light*. He wrote:

> The Holy Blessed One, for the sake of His compassion and His loving-kindness, prepared one day in the year for Israel to purify themselves . . . and called it Yom Kippur [in Hebrew, *Yom HaKippurim*, which has a plural ending], in plural language because these two sefirot unify on that day: the sefirah of *Binah* [the supernal mother] and the sefirah of *Malchut* [the daughter, or the lower mother] . . . And when these two sefirot join together to reverse by their merit the dross of Israel, to purify them, the day is called *Yom HaKippurim* . . . As these two sefirot attune, each facing the other, in the secret of the Supernal Mother and the Lower Mother, they occupy themselves on this day with the purification of Israel, therefore [presumably heterosexual] sex is forbidden on Yom Kippur, although it is permitted on Shabbat and festivals . . .[12]

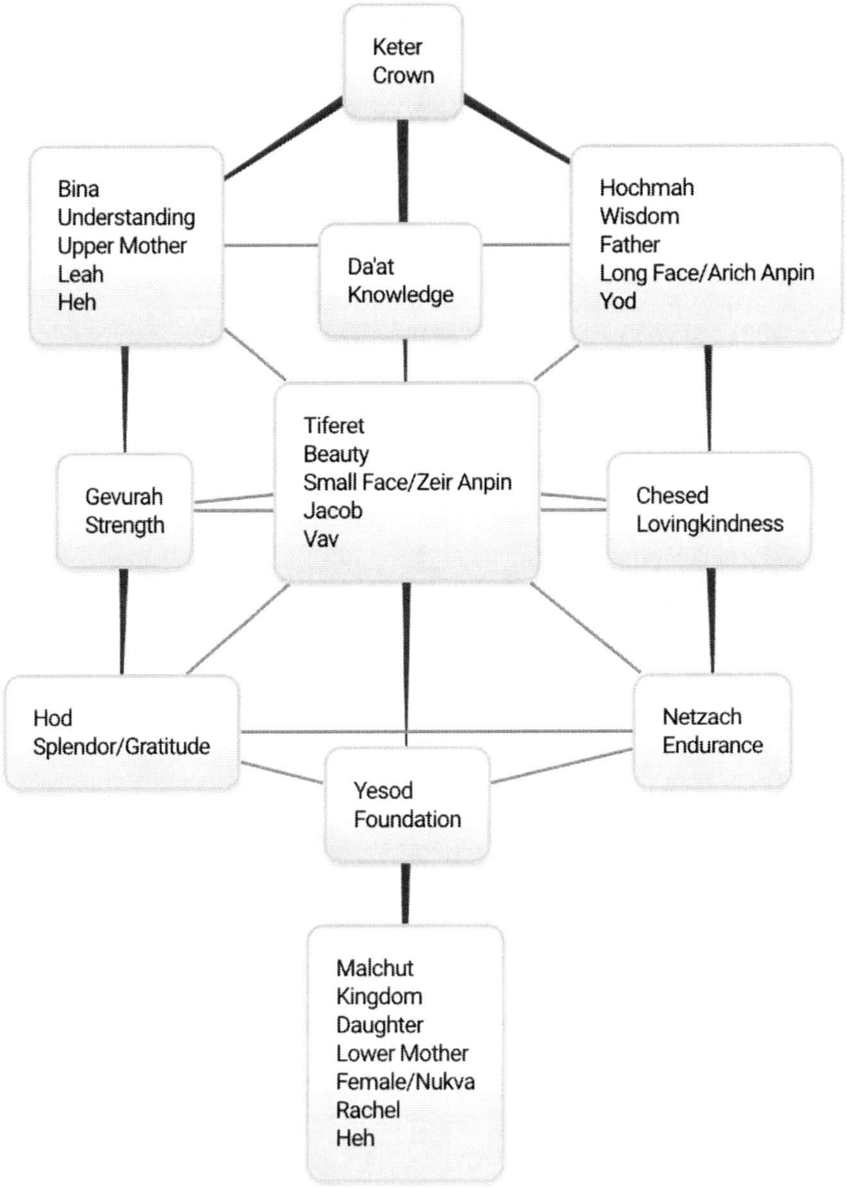

Figure 10.1. Tree of Life diagram, based on Lurianic Kabbalah, including terms used in this article.
Diagram by the author.

How striking! Yom Kippur, the holiest day of the Jewish year, is the day on which the two mothers of the Tree of Life engage together. Here, the entities that are unifying are imagined as female, so whatever energy flows between them need not place one of them on one end of a gender polarity and the other on the "opposite" end. However, as will become clear later, they also are not the "same" gender. In any case, the joining together of these two female aspects accomplishes the goal of purifying the people during the holiest time of the year. This is a truly lovely take on the erotic dynamics of Kabbalah.

COUPLING WITHIN PHENOMENA

Rachel and Leah

Among the deepest and most evocative teachings about gender in Jewish sacred texts are those involving the dynamics of the biblical matriarchs Rachel and Leah, two of the most central ancestors of the Jewish people. We will explore their story through some layers of text, approaching it as a phenomenon. We start in the middle of the story, in Genesis 30:14–24. In this passage, the sisters Rachel and Leah, who are both married to Jacob, bargain with each other for the mandrakes that were found by Leah's son Reuben. The mandrakes are valued as an aid to fertility. In the trade they arrange, Leah gets to spend the night with Jacob, who prefers Rachel. Rachel, who has been infertile, gets the mandrakes. Leah, after her night with Jacob, conceives and gives birth to Issachar. Rachel then conceives and gives birth to Joseph. Who, then, is responsible for each of these children's formation? Reuben? Rachel? Leah? Rachel and Leah together? Jacob? God? The mandrakes themselves?[13] The *phenomenon* that brings each of these children to birth extends far beyond the male-female duos whose egg and sperm were involved.

Phenomena from a Physics Perspective

When I use the term *phenomenon*, I am referring specifically to the definitions of the term used by physicist Niels Bohr, and developed by philosopher, physicist, and queer theorist Karen Barad, in their theory "agential realism."[14] Getting a clear sense of this concept takes us now on a detour into the philosophy of physics. Barad defines phenomena like this: "Phenomena are entanglements of spacetimemattering, not in the colloquial sense of a connection or intertwining of individual entities, but rather in the technical sense of 'quantum entanglements,' which are the (ontological) inseparability of agentially intra-acting 'components.'"[15]

An example Barad uses to illustrate these concepts is that of measurement of the position and momentum of a particle.[16] Quantum physics brings to our attention the problem that it is impossible to measure both of these variables at the same time, because the measurement of *position* requires an apparatus with fixed parts (for example, a photographic plate supported by a fixed platform, which doesn't move while a photon hits the plate, whereas measuring *momentum* requires an apparatus with movable parts—a platform which moves as the photon hits it, thus enabling it to measure momentum. Barad explains that Bohr concludes that on this basis "there is no unambiguous way to differentiate between the 'object' and the 'agencies of observation.'"[17] In the case of a position measurement, the photon is part of the agencies of measurement, whereas in the case of the measurement of momentum, the photon is the object of measurement. Therefore,

> *The boundary between the "object of observation" and the "agencies of observation" is indeterminate in the absence of a specific physical arrangement of the apparatus.* What constitutes the object of observation and what constitutes the agencies of observation are determinable only on the condition that the measurement apparatus is specified. *The apparatus enacts a cut delineating the object from the agencies of observation.* Clearly, then . . . *observations do not refer to properties of observation-independent objects (since they don't preexist as such).*[18]

Barad thus explains the notion of phenomena (emphasis mine):

> Since individually determinate entities do not exist, measurements do not entail an interaction between separate entities; rather, determinate entities emerge from their intra-action. I introduce the term "intra-action" in recognition of their ontological inseparability, in contrast to the usual "interaction," which relies on a metaphysics of individualism (in particular, the prior existence of separately determinate entities). *A phenomenon is a specific intra-action of an "object" and the "measuring agencies"; the object and the measuring agencies emerge from, rather than precede, the intra-action that produces them. Crucially, then, we should understand phenomena not as objects-in-themselves, or as perceived objects (in the Kantian or phenomenological sense), but as specific intra-actions.*[19]

Understanding the arising of children in the mandrake story as intra-actions within a phenomenon makes it clear that there are no inherently determinate pre-existing entities, such as an individual named Rachel, or Leah, or Jacob, or their children, but rather these "individuals," and the rest of the phenomenon of which they are a part, arise out of their intra-actions, which include the entire apparatus of which they are a part.

And there is no determinate end to the boundaries of an apparatus. Barad illustrates this by telling the story of the 1922 Stern-Gerlach experiment, which attempted to prove space quantization (the fact that electrons only inhabit specific orbits, and can't be in between those locations).[20] Their measurement apparatus used beams of silver atoms to make marks on a photographic plate. Their labors seemed to be showing nothing, until Gerlach handed the photographic plate to Stern, and as he breathed on it, the image they were seeking emerged. Stern was an assistant professor with a low salary, so he smoked cheap cigars, which contained a lot of sulfur. The sulfur combined with the silver to make the image visible. Barad explains:

> Apparatuses are not static laboratory setups but a dynamic set of open-ended practices, iteratively refined and reconfigured . . . a cigar is among the significant materials that are relevant to the operation and success of the experiment . . . Not any cigar will do. Indeed, the cigar is a "condensation"—a "nodal point" as it were—of the workings of other apparatuses, including class, nationalism, economics, and gender, all of which are a part of this Stern-Gerlach apparatus . . . in this case, material practices that contributed to the production of gendered individuals also contributed to the materialization of this particular scientific result . . . : "objects" and "subjects" are coproduced through specific kinds of material-discursive practices. Stern's gendered and classed performance of masculinity (e.g., through his cigar smoking) mattered . . . This example not only illustrates the dynamic nature of scientific practices and the lack of a determinate outside boundary to the apparatus but also clearly suggests that humans enter not as fully formed, preexisting subjects but as subjects intra-actively co-constituted through the material-discursive practices that they engage in.[21]

So in the phenomenon of Rachel, Leah, and Jacob's family, one might explore the extent of the apparatus of which they are a part. One might want to explore questions of "class, nationalism, economics, and gender." And clearly, this web of connection that produced children is not a simple binary process. One need not imagine the process of creation—divine, human, or any other—as something involving solely two determinately individual, determinately sexed beings—in this case, Jacob and Rachel or Jacob and Leah. It is clear that Rachel and Leah (as well as the mandrakes) were more in control of the reproductive activity of the family than Jacob was, so the common kabbalistic image for creative flow involving Jacob/*Tiferet* giving,[22] and Rachel/*Malchut* receiving, is clearly simplistic. But even more—understanding the mandrake story as a phenomenon, all the entities that appear to be separate "individuals" can be seen as "iteratively refined and reconfigured," arising through a process of being "intra-actively co-constituted through the material-discursive practices that they engage in." This expansion of awareness of relationality opens up new ways of looking at what it means to be part of the process of creating life.

Rachel and Leah as the Phenomenon of Breasts

In Jewish sacred traditions, Rachel and Leah are often imagined as the two breasts,[23] flowing nourishment and blessing into the world. They are associated with the upper and lower Mother *partzufim* of the Tree of Life, and we will discuss their kabbalistic unification shortly. But of note here is that the word for mandrakes, *dudaim*, is taken in midrashic tradition as related to the root for breast, *dad.* The mandrakes, then, are a reminder that the two sisters are part of one phenomenon—the two breasts may appear separate, and they certainly are distinguishable, but they are nonetheless clearly part of the larger apparatus of a body. As such, what occurs between them is intra-action, out of which their identities arise. [24]

As Ellen Davina Haskell has explored so beautifully in her book, *Suckling at My Mother's Breasts*,[25] the image of breasts and breastfeeding is used in Jewish tradition to evoke the giving of everflow (*shefa*) and blessing to the world. The flow of milk is also identified with the transmission of Torah and Jewish identity.[26] One of the most central Hebrew prayers, the *Aleinu*, contains the phrase "to repair the world with the kingdom of *Shaddai,*" a name for the divine that is often translated into English as "Almighty." It is etymologically related to a word for mountain, but homiletically and homologically refers to *shaddayim*, another word for breasts.[27] A Lurianic[28] prayerbook suggests that this prayer phrase is referring to Rachel and Leah.[29] That is, to repair the world with the kingdom of *Shaddai* means to repair the world with the divine Name that implies breasts, and evokes the aspects of Rachel and Leah. World repair happens by means of the power of the Mother, whose breasts remind us that those entities that appear separate are actually part of a larger phenomenon. The flow of blessing into the world comes from the two female *partzufim*, the Upper and Lower Mothers, Binah and Malchut, Leah and Rachel. That is, the givers of flow are both female. [30]

The Indeterminacy of Rachel and Leah

Now we return to the Torah story of Rachel and Leah. Genesis 29:16–28 describes what happened before they were married to Jacob. Jacob loved Rachel, but Rachel's father, Lavan, tricked Jacob into marrying Rachel's older sister, Leah, instead. After they spend their wedding night together, Jacob discovers the switch, and arranges to marry Rachel as well. A variety of Midrashim[31] attempt to solve the problem of why Jacob took so long to figure out what was happening, and explore what was going on between Rachel and Leah. *Genesis Rabbah* 70:19 tells it like this: "The whole of that night he called her 'Rachel,' and she answered him. In the morning, however, 'behold it was Leah.'"[32] That is, Leah replies as if she were Rachel, not herself.

Lamentations Rabbah, Prologue 24, relates that at the time of the destruction of the Temple, and the beginning of the Babylonian exile, the patriarchs Abraham, Isaac, and Jacob, as well as Moses, were brought before the Holy Blessed One. Each made an argument justifying why God should have compassion on the children of Israel, based on his own meritorious acts, but none convinced the Holy One. Finally, Rachel burst forth, and gave an argument that convinced the One to say "For your sake, Rachel I will restore Israel." Here is part of what the Midrash tells us Rachel said:

> Sovereign of the Universe . . . my father planned to substitute another for me to wed my husband for the sake of my sister. It was very hard for me, because the plot was known to me and I disclosed it to my husband; and I gave him a sign whereby he could distinguish between me and my sister, so that my father should not be able to make the substitution. After that I relented, suppressed my desire, and had pity upon my sister that she should not be exposed to shame. In the evening they substituted my sister for me with my husband, and I delivered over to my sister all the signs that I had arranged with my husband so that he should think that she was Rachel. More than that, I went beneath the bed upon which he lay with my sister; and when he spoke to her she remained silent and I made all the replies in order that he should not recognize my sister's voice. I did her a kindness, was not jealous of her, and did not expose her to shame . . . [33]

Whereas in *Genesis Rabbah*, Leah spoke as if she were Rachel, here *Lamentations Rabbah* explains that Rachel replied as if she were Leah, who was herself pretending to be Rachel. In these stories, the two sisters' identities are what quantum physics would call indeterminate. By indeterminate, I don't mean uncertain.[34] According to some perspectives, there are measurements that are uncertain, because they can't be fully measured. This is the view that 20th century physicist Werner Heisenberg expressed in his famous uncertainty principle—the idea that, for example, one can't measure the position and momentum of a particle simultaneously, because when one of these things is measured, the observer disturbs the other variable, thus making it impossible to know both at the same time. But returning to what we discussed earlier, Niels Bohr had a different understanding of what the measurement problem entailed: As Barad explains it, for Bohr "observations do not refer to properties of observation-independent objects (since they don't preexist as such)." That is, the qualities of position and momentum are not uncertain, in that one is disturbed in measuring the other, but rather they are indeterminate—not that one doesn't know the full measurement of something, but that the thing *has no* pre-existing determinate measurement to know. [35] The same idea that holds for the position and momentum of a particle also holds for determining whether an entity is a particle or a wave.[36] That is, not only the properties of something (i.e., an entity's position or momentum), but its very nature—whether it is a particle or a wave—is inherently indeterminate in the absence of a particular apparatus. Barad concludes:

> *So the very nature of the entity—its ontology—changes (or rather becomes differently determinate) depending on the experimental apparatus used to determine its nature.* For Bohr . . . the objective referent for concepts, like "wave" and "particle," is not a determinately bounded object with inherent characteristics . . . but rather what he called a *phenomenon*—the entanglement/inseparability of "object" and "apparatus" (which do not preexist the experiment but rather emerge from it). [37]

It is Rachel's willingness to let go of her individual identity, her "nature," to not be her "self," or to let Leah be her, or to let the two of them be indeterminate, to be clearly part of the same phenomenon, that led to redemption.[38] What would the world be like if actions arose from awareness of "ongoing iteratively intra-active reconfiguring" within larger phenomena, rather than from imagining the world as made up of determinately separate small individuals? Perhaps that is redemption.

GENDER INDETERMINACY

As with any aspect of identity, or any measurable quality, gender can also be seen as inherently indeterminate. Gendering beings arise through iterative intra-actions, enacting specific agential cuts. Butler writes that: " . . . gendering is, among other things, the differentiating relations by which speaking subjects come into being . . . the 'I' neither precedes nor follows the process of this gendering, but emerges only within the matrix of gender relations themselves."[39] Barad explains: "Gendering, Butler argues, is a temporal process that operates through the reiteration of norms. In other words, Butler is saying that gender is not an inherent feature of individuals, some core essence that is variously expressed through acts, gestures, and enactments, but an iterated doing through which subjects come into being."[40]

Gender and Other Indeterminacies of Leah and Rachel

The arising of identity and gender out of relational matrices can be seen at play throughout Jewish mystical texts. The Zohar comments on Genesis

29:31,[41] "And Adonai saw that Leah was hated," by using a verse from Psalm 113 (Ps 113:9): " He settles *aqeret habayit*, the barren woman in her home, as a joyous mother of children" saying, "He settles *aqeret habayit*—Rachel, *iqqara deveita*, essence of the house. As a joyous mother of children—Leah." There is a lovely Hebrew/Aramaic play on words with the root *ayin quf resh*, taking *aqeret* "infertile woman" according to the plain meaning, and understanding it to mean another word that shares the same root—*iqqara*, "essence," thus establishing first of all that whatever associations one might have with infertility, it does not mean that an infertile woman is not essential. The Zohar goes on to say that this verse means that God is transforming Rachel, *aqeret habayit*, the infertile or essential one of the household, into Leah, *em habanim s'meicha*, a happy mother of children. The text goes on to associate Rachel and Leah, respectively, with sabbatical year (*Shmitta*) and Jubilee (*Yovel*), revealed lower world, and concealed higher worlds, *Attah*/You and *Hu*/He, and implicitly, *Binah* and *Malchut*. Divine power shifts Rachel into the place of Leah. Not that Rachel becomes completely undifferentiable from Leah, but that they are clearly seen to be part of the same phenomenon, with boundaries that are ontologically indeterminate. And similarly, the indeterminacy of Rachel and Leah is overlaid with all the other levels of association listed by the Zohar—the seven year sabbatical and the fifty year Jubilee cycle; revealed and concealed worlds; You (second person) and He (third person); what might be imagined as inherently and determinately separate entities are shown to have indeterminate boundaries, identities that arise out of their intra-action within the phenomena of which they are part.

Another interesting implication of this Zohar passage is the association of Leah with the Hebrew name "*Hu*," the English word "he." Zohar I 154b examines an oddity of language in Genesis 30:16. In the English, it reads: "And Jacob came from the field in the evening, and Leah went out to meet him, and said, You must come in to me; for I have hired you with my son's mandrakes. And he lay with her that night." But the phrase "that night" in Hebrew is unusual—rather than the usual way to say "that night," "*balaila hahu*," it reads "*balaila hu*," leaving out the prefix "*ha*" on "*hu*" which would have made the word meaning "that," so that the plain meaning instead is "in the night—he." The Zohar's implication is that (similar to the surprise Jacob felt on their wedding night, in Genesis 29:25), suddenly in the night—*balaila—hu!*—he! During their night together, Jacob somehow experienced Leah as "he." The Zohar goes on to associate *Hu* with *Binah*, and explain that the reason Jacob hated Leah is that, just as *Binah* is the mother of *Tiferet*, which is associated with Jacob, Leah seemed like a mother to Jacob, so therefore she was not appealing to him as a partner. But the passage also may imply that what didn't work for Jacob was the maleness of Leah/*Binah*.[42]

The Zohar adds "how can you say 'that Leah was hated'? . . . Jubilee is always a concealed world, none of whose matters are revealed; so all its actions were concealed from Jacob . . . All was concealed from Jacob, for the higher world is completely unrevealed . . . Jacob, aspiring, did not cleave to the concealed, rather to the revealed." [43] The implication is that Jacob didn't relate to Leah because he didn't understand her—she was at too high a level for him. In the relation between Jacob and Leah the stereotypical gender dynamic is turned upside down.

It is clear that the relation between Leah and Jacob, or *Binah* and *Tiferet* in the language of Zohar (or *Ima* and *Zeir Anpin* in Lurianic language), is one of active female and passive male. Rabbi Yehuda Ashlag, in *HaSulam*, his Lurianic commentary on Zohar, comments on Zohar I 154a:

> **A happy mother of children: The Jubilee**, which is Binah; **all happiness, and all joy in all the worlds . . . are dependent on her**, because the female (nukva) of Zeir Anpin has nothing on her own, only what Zeir Anpin receives from Binah, and directs the flow of toward her, and then she directs the flow to all the worlds. Behold, **all happiness that is in all the worlds** . . . comes from Binah.[44]

So, why would one think on some essential level that the male is giver and the female receiver? It is necessary to rethink the whole system of gendered associations.

SHIFTING THE CENTRAL PAIR TO THE MOTHER AND HER CHILDREN

Another approach to the dynamics of creation and creating is evident in texts that focus on the relation of the Mother and her children. Zohar II 104b-105a[45], associates *Mi*, 'who,' with *Binah*, and *elleh*, 'these' with the twelve tribes of Israel. In this passage, *elleh* exists to spread the name *Mi*, which in turn will encompass Israel in the divine name *Elohim*, the letters of which are formed by the joining together of the *Mi* and the *Elleh*. (*Elohim* is the name of God used in the Genesis creation story.) But, this joining is threatened. As the text explains it:

> For in exile, *mi*, Who, has withdrawn above—Mother, as it were, away from children and the children have fallen. And the name that was complete—that supernal, primordial, grand name—has fallen. For this we pray and sanctify in synagogue, that this name may be restored as it was: "May His great name be enhanced and sanctified!" Who is "His name"? That "great" one, first of all. For it has no structure without us: *Mi*, who, is never built up without *Elleh*, these. So,

> at that time, *Mi elleh*, Who (are) these, that fly like a cloud? The whole world will see that supernal name has been restored to perfection.[46]

This text is advocating for a focus on the unification of the name *Elohim,* by unifying the *mi* and the *elleh,* mother Binah and her children. This will lead to the world being whole, redeemed, and "fragrant" as the Zohar puts it. The text explains that the highest and most spiritually powerful response in Jewish liturgy—*yehei shmei rabbah mevorach*—May His great name be blessed—is referring to this process. Lurianic Kabbalah often includes *kavvanot*, intentions to be said before engaging in a holy act, that state that the act is being done with the intention of unifying the *Kadosh Baruch Hu*[47] and the *Shechinah,*[48] but this Zohar asserts the centrality of the unification of the Mother and her children.[49]

THE FLUIDITY OF IDENTITY AS IT ARISES OUT OF TEXT

Throughout the Zohar (and other Jewish mystical texts to a lesser degree), associations of words, sefirot, concepts, and genders shift between passages, and even within them. For example, Zohar I: 2a at one point associates *Elleh* with Leah, as the words are anagrams of each other, and also associates *Elleh*/ Leah with *Binah*, associating *Mi* with *Hochmah*, but in another part of the same passage, it associates *Mi* with *Binah, Elleh* with the lower six sefirot, and *Mah* with Malchut. In that same passage, Malchut borrows Binah's male garments and then appears as Elohim, the combination of Binah and the lower six sefirot. What are seen as individual entities, and what are not, what is entangled with what, and in what ways, has to do with the phenomena and apparatuses involved—"the entanglement—the ontological inseparability—of intra-acting agencies."[50]

THE INDETERMINACY OF INSIDE AND OUTSIDE

Lurianic Kabbalah goes into more detail about the structure of the relationship between Rachel/*Malchut* and Leah/*Binah*, or in Lurianic terms, *Nukva* (Female) and *Ima* (Mother). The 18th century kabbalist R. Moshe Chaim Luzzatto explains that according to Lurianic tradition,[51] Leah is the back of *Ima*, and the inner aspect of *Nukva*, Rachel is the front of *Ima* and the outer aspect of *Nukva*.[52] In the texts we have already discussed, it was clear that Rachel and Leah have indeterminate boundaries and identities, and at times it appears that they can become one another; if so, it seems that Luzzatto could have been describing the shape of a Klein bottle. A Klein bottle is a topological structure in which the inside becomes the outside, becomes the inside. Similar to the more familiar Mobius strip, a surface with one continuous side formed by joining the ends of a rectangular strip after twisting one end 180°, a Klein bottle is a closed surface with only one side, formed by passing one end of a tube through the side of the tube and joining it to the other end. It is possible to make a close approximation of a Klein bottle out of two Mobius strips zipped together along their edge. One could imagine *Binah* and *Malchut* as a pair of Mobius strips, each one with Leah on one side and Rachel on the other, becoming the other as one follows the strip around, and then when *Binah* and *Malchut* are zipped together, unified, so to speak, the result is a Klein bottle that could describe the structure Luzzatto draws. That is, if *Malchut* and *Binah*, daughter and Mother, or two sisters, Rachel and Leah, are unified—that is, seen as part of the same apparatus or phenomenon— the result is a Klein bottle. Again, the dynamic has shifted from *Zeir Anpin*[53] giving to *Nukva* as the primary focus, where the goal is the unification of the Holy Blessed One and Shekhina. The influx from *Zeir Anpin*, which comes from *Ima*, could be seen as being for the sake of unification, or perhaps interplay, of the two female aspects, *Ima*/*Binah*/Leah and *Nukva*/*Malchut*/Rachel. The sisters, who are also mother and daughter, front and back, inner and outer, in an endlessly shifting dynamic, are the central pair.

It then becomes clear that the whole process of inside and outside can be understood like this, one becoming the other. In Lurianic Kabbalah (which describes the structure of the process of emanation of creation that started with transcendent nothingness, and chained down into the material world), there is a series of switches between the higher level being on the inside, and the lower on the outside, and the higher level being outside, and the lower level inside. The entire system is a Klein bottle, or rather, a series of Klein bottles.[54]

REDEMPTION AS CONSCIOUS ACCEPTANCE OF INDETERMINACY

Joseph Gikatilla, in *Sha'arei Orah (Gates of Light)*, associates *Binah* with the divine name that is spelled *YHVH* but vocalized as *Elohim*.[55] Writing about the need to connect the upper and lower Shechinahs (that is, *Binah* and *Malchut*), Gikatilla writes:

> Those therefore who know how to please their Creator know how to repair the way to the SHeCHINaH, to bring her back to her place and to repair the channels that have been ruined; then the upper SHeCHINaH will bestow her blessings upon the upper beings, which allows the other Spheres to fill the lower SHeCHINaH, and that allows her to return to her place to bring forth blessing to

> the world . . . this notion is in the verse which applies to both of them: "And you know on this day and you will take it to your heart that YHVH is ELoHIM in the heavens above . . ."—this is referring to the lower SHeCHINaH[56] (Deuteronomy 4:39). If you would say from this that there are two separate realms, God forbid, know that they are only one—"For YHVH, He is the ELoHIM in the heavens above and in the earth below, there is no other."[57]

Gikatilla teaches here that the two mother aspects of the divine, *Binah* and *Malchut*, or the upper and lower Shechinah, are called by two central of divine names, YHVH and Elohim—and that the book of Deuteronomy (in a verse quoted in a prayer that is said multiple times a day), is making it clear that even though they seem like two separate entities, they are actually part of the same divine phenomenon. The realization, and enactment, that these two levels, *Binah* and *Malchut*, are only *agentially separable* is what redeems the world, because reuniting these two aspects, or recognizing they are not essentially separate, is what allows the everflow of life and blessing to pour forth into the world.

In a section following the passage from *Gates of Light* referred to earlier, (about the unification of *Binah* and *Malchut* on Yom Kippur), Gikatilla comments on Isaiah 1:18, "If your sins are like crimson, they will be bleached as the snow," saying

> The upper Sphere [that is, Binah], is called LeBaNON [same root as the word for white], and the other [Malchut] is seen to be dressed in crimson, so Israel must transform its crimson garments into white ones, and that is why it is called YOM HaKiPURIM, and this is the essence of the crimson which was bleached as an atonement in Temple ritual, for the Sphere BINaH appeared to enlighten the Sphere MaLHUT . . . [58]

What brings atonement at Yom Kippur, described by tradition as scarlet thread becoming white, can be understood as *Malchut* becoming *Binah*—or, in other terms, Rachel becoming Leah. This is what brings redemption.[59]

The redemptive act is Rachel's love of Leah, and her willingness to be Leah, and let Leah be her. Becoming Leah means becoming the concealed, the genderqueer, the mother, the back, the inner, the creator, the primordial name. It means accepting indeterminacy. It means recognizing that the whole purpose of creation is what is taking place within the Creator, within Elohim, a reuniting of Rachel and Leah, a recognition that the separation between Mother and child is purely an agential cut, as are all other separations.

Not that there is no differentiation to be found—differentiation is what makes creation possible, what allows for giving and receiving, what allows for play. And most certainly, gender play is among the most joyous aspects of creation. Yet recognizing that there is no need to focus solely on one particular aspect of that play, a vertical heterosexual union, but rather a multitude of unifications can be recognized, celebrated, and enacted—those between the Mother and her children, between the two mothers, the two sisters, and many more—adds so much to the richness of life, of love, of identity, and ultimately of experience of divine presence in the world and in life.[60]

WORKS CITED

Afilalo, Raphaël, and Moshe Ḥayyim Luzzatto. *The Kabbalah of the Ari Zal According to the Ramhal: Moshe Hayim Luzzatto: Kelalot Ha-Ilan Ha-Ḳadosh*. Montreal: Kabbalah Editions, 2004.

Barad, Karen. *Meeting the Universe Halfway: Quantum Physics and the Entanglement of Matter and Meaning*. Durham, NC: Duke University Press, 2007.

———. "Nature's Queer Performativity." *Kvinder, Køn and Forskning* 12 (2012): 25–53.

———. "Posthumanist Performativity: Toward an Understanding of How Matter Comes to Matter." In *Signs: Journal of Women in Culture and Society* 28, no. 3, 801–831. Chicago: University of Chicago Press, 2003.

———. "TransMaterialities: Trans*/Matter/Realities and Queer Political Imaginings." *GLQ: A Journal of Lesbian and Gay Studies* 21, no. 2–3 (2015): 387–422. doi:10.1215/10642684–2843239.

Barad, Karen, and Adam Kleinman. "Intra-actions." *Mousse Magazine*, June 1, 2012, 76–81.

Biale, David. "The God with Breasts: El Shaddai in the Bible." *History of Religions* 21, no. 3 (1982): 240–56.

Boyarin, Daniel. *Unheroic Conduct: the Rise of Heterosexuality and the Invention of the Jewish Man*. Berkeley: University of California Press, 1997.

Butler, Judith. *Bodies That Matter: On the Discursive Limits of "Sex."* New York: Routledge, 1993.

———. *Gender Trouble Feminism and the Subversion of Identity*. New York: Routledge, 1999.

Drob, Sanford. "The Sefirot." *New Kabbalah*. Accessed February 14, 2017. http://www.newkabbalah.com/sefirot.html.

Gikatilla, Joseph. *Sefer Sha'arei Orah: Hotza'at Chadashah Umetukenet*. Edited by Yeroham Bakar. Jerusalem: 2004 ,חמו"ל.

———. *Gates Of Light = Sha'Are Orah*. San Francisco: HarperCollins, 1994.

Haskell, Ellen Davina. *Suckling at My Mother's Breasts: The Image of a Nursing God in Jewish Mysticism*. Albany: SUNY Press, 2012.

Hellner-Eshed, Melila, and Nathan Wolski. *A River Flows from Eden: The Language of Mystical Experience in the Zohar*. Stanford, CA: Stanford University Press, 2009.

Luzzatto, Moshe Ḥayyim. *138 Openings of Wisdom = 138 Pitḥei Ḥokhmah*. Jerusalem: Azamra Institute, 2005.

Matt, Daniel C. *The Zohar*. Pritzker ed. Stanford, CA: Stanford University Press, 2009.

Patai, Raphael. *The Hebrew Goddess*. 3rd enl. ed. Detroit, MI: Wayne State University Press, 1990.

Schachter-Shalomi, Zalman. *Yishmru Daat: Hasidic Teachings of the Fourth Turning* (Hebrew Edition). Boulder, CO: Albion-Andalus Books, 2013.

Schneider, Sarah. *Kabbalistic Writings on the Nature of Masculine and Feminine*. Northvale, NJ: Jason Aronson, 2001.

Shaharabani, Yehudah Ben G'org'i Eliyah. *'Oz ha-tefilah: sidur tefilat Shaḥarit: 'im leḳeṭ ḳitsur Ḳayanot ha-Ari . . .* Petaḥ-Tiḳyah: Y. Shaharabani, 1998.

Sperling, Harry. *The Zohar*. 2nd ed. London: The Soncino Press, 1984.

Yohai, Simon. *Sefer HaZohar Im Payrush HaSulam*. Jerusalem: Yeshivat Kol Yehuda, 1991.

NOTES

1. Karen Barad, "TransMaterialities: Trans*/Matter/Realities and Queer Political Imaginings," *GLQ: A Journal of Lesbian and Gay Studies* 21, no. 2–3 (2015): doi:10.1215/106l42684–2843239.

2. Ya'akov Yosef of Polonoye, *Toldot Ya'akov Yosef*, quoted by Raphael Patai, *The Hebrew Goddess* (Detroit: Wayne State University Press, 1990).

3. See the work of Daniel Boyarin, particularly *Unheroic Conduct: The Rise of Heterosexuality and the Invention of the Jewish Man.*

4. The term Midrash refers to the rabbinic tradition of stories and legal interpretations based on oral traditions related to the Hebrew Bible. Some use the term only to apply to ancient and medieval rabbinic texts, while others include what might be called "modern midrash," sacred stories that are still being crafted.

5. Sarah Schneider, *Kabbalistic Writings on the Nature of Masculine and Feminine* (Northvale, NJ: Jason Aronson, 2001), 31.

6. In addition to what is discussed here, consider the diversity of biological sex of intersex people.

7. Judith Butler, *Bodies That Matter: On the Discursive Limits of "Sex"* (London and New York: Routledge, 1993), 1–2.

8. Butler here focuses on the ways in which sex is formed by discursive practices. I would argue, along with Barad, that sex and gender are co-constituted.

9. Zalman Schachter-Shalomi, *Yishmru Daat: Hasidic Teachings of the Fourth Turning* (Boulder: Albion-Andalus Books, 2013). Emphasis in the original.

10. Jeremiah 31:21.

11. Sanford Drob, www.newkabbalah.com gives this definition: "The *Sefirot* (singular *Sefirah*) which are almost always conceived to be ten in number, are the building blocks of creation, the archetypes of existence, the traits of God, and the primary values of the world." Please refer to the diagram of the Tree of Life on page 203 to see some of the traditional associations of the sefirot that I will refer to in this chapter.

12. Joseph Gikatilla, *Sha'arei Orah* (Jerusalem: חמו"ל, 2004), 346–347. Translation mine. Thank you to Nathaniel Berman for pointing this text out to me many years ago.

13. The Hebrew word for mandrakes, *dudaim*, shares a root with words for breasts, beloved, cauldron, and an epithet for G-d. How could they not be involved in the process?

14. Thank you to Karen Barad for the many shared flights of mind in which we read these stories through one another.

15. Karen Barad, "Nature's Queer Performativity," *Kvinder, Køn and Forskning* 12 (2012): 32.

16. Karen Barad, *Meeting the Universe Halfway* (Durham: Duke University Press, 2007).

17. Barad, *Meeting the Universe Halfway*, 114.

18. Barad, *Meeting the Universe Halfway*, 114. Emphasis in the original.

19. Barad, *Meeting the Universe Halfway*, 128. Emphasis mine, except for first emphasized phrase, which is in the original.

20. Barad, *Meeting the Universe Halfway*, 161–68.

21. Barad, *Meeting the Universe Halfway*, 167–68.

22. In the Kabbalistic Tree of Life, *Tiferet* (Beauty) is the sefirah that is generally associated with the *Kadosh Baruch Hu*, or Holy Blessed One, who in unifying with the lowest sphere *Malchut* (Kingdom) associated with the female immanent presence of the divine, brings life into the world. These aspects are associated with Jacob and Rachel, and in Lurianic Kabbalah with the *partzufim* of son and daughter, also called "small face" (*Zeir Anpin*) and "female" (*Nukva*). A *partzuf* (plural *partzufim*), literally "persona" is a configuration of a sefirah or sefirot into an interface, which acts as a filter and connector of the flow of divine emanation from the infinite source to the created world. Please refer to the chart on page 203 to see how these qualities exist within a larger schema.

23. See Joseph Gikatilla, *Sha'arei Orah, English in Gates of Light*, The Eight Gate, the Third Sphere, p. 303, Avi Weinstein translator; Zohar II 22a and Daniel Matt fns, *Siddur Oz HaTefillah*, R. Yehudah Shaharabani on *b'Malchut Shaddai* in *Aleinu.*

24. A similar image appears in the Torah's description of the *cheruvim*, angels, on the ark of the covenant in the Holy of Holies (the holiest place in Jewish tradition): the two of them were to be hammered out of one piece of gold—like the breasts, a teaching that what appear to be separate entities are part of a larger phenomenon. For more on this topic, see my upcoming article.

25. Ellen Davina Haskell, *Suckling at My Mother's Breasts: The Image of a Nursing God in Jewish Mysticism* (Albany: SUNY Press, 2012).

26. See Haskell, *Suckling at My Mother's Breasts: The Image of a Nursing God in Jewish Mysticism.*

27. See David Biale, "The God with Breasts: El Shaddai in the Bible," *History of Religions* 21, no. 3 (1982): 240–256 for more on this topic.

28. Lurianic Kabbalah, developed by Rabbi Isaac Luria and his followers in the 16th century, was a transformative and complex development inspired by the Zohar, which has had tremendous influence on Kabbalah since then.

29. Yehudah Ben G'org'i Eliyah Shaharabani, *'Oz ha-tefilah: Siddur Tefilat Shaḥarit: 'im leḳeṭ ḳitsur Ḳavanot ha-Ari* (Petach Tikvah: Y. Shaharabani, 1998).

30. However, interestingly, Haskell points out many midrashim where breast-feeding is associated with the patriarchs, queering the image; one midrash imagines Mordechai breastfeeding his niece Esther, others picture Moses and Aaron, and Torah scholars in general, as breasts.

31. In addition to the pieces mentioned below, also see *Baba Batra* 123a.

32. Genesis 29:25.

33. Adapted from the Soncino Classics translation.

34. Many thanks to Karen Barad for drawing out this crucial distinction in *Meeting the Universe Halfway* and elsewhere.

35. See Barad, *Meeting the Universe Halfway,* 114, and elsewhere.

36. See ibid, chapter 3 for a full discussion of this idea.

37. Barad, *Nature's Queer Performativity*, 42. Emphasis in the original.

38. Redemption is one of the most central concepts in Judaism, and can be understood to mean communal liberation from suffering.

39. Butler, *Bodies that Matter,* 7.

40. Barad, *Meeting the Universe Halfway*, 57.

41. Zohar I 154a-b, translation Daniel Matt.

42. This is supported by a variety of texts that discuss *Binah*'s maleness. For example, see Zohar I, 1b-2a.

43. Zohar I 154a-b, translation Daniel Matt.

44. R. Yehuda Ashlag, Sefer Zohar Peirush HaSulam, Vol 5, 65, translation mine. Bold reflects bold in the original.

45. As well as other sections of the Zohar, including I: 1–2.

46. Zohar II 105a, translation Daniel Matt.

47. The Holy Blessed One, often but not always associated with *Tiferet* and Jacob.

48. Associated with *Malchut*, and sometimes Rachel.

49. This may reverberate in the text referred to earlier by Rabbi Shachter-Shalomi, in which the lovers are children who suckle together at the breasts of the mother. In that image, the lovers are receiving flow together, and ultimately, the most primary unification is between each of the children and the mother.

50. "Intra-actions," an interview with Karen Barad, PhD by Adam Kleinman in *Mousse Magazine*, 2012, 77. This use of the term entanglement follows the technical quantum sense and is not restricted to its colloquial usage.

51. Moshe Chaim Luzzatto, *Klach Pitchei Hochmah, 138 Openings of Wisdom*, Translated by Avraham Yehoshua Greenbaum (Jerusalem: Azamra Institute, 2005).

52. Please refer to the chart to see where these are situated on the Tree of Life.

53. The Lurianic term for the *partzuf* that includes *Tiferet*, often associated with Jacob.

54. The biggest of these Klein bottles includes the two paradigms for the structure of emanation, the *Iggulim* and the *Yosher,* but there are many others, which will be the topic of another paper.

55. Gikatilla, *Gates of Light*, Beginning of The Eight Gate, The Third Sphere. In English, 28.

56. This verse from Deuteronomy is part of the same prayer for redemption mentioned earlier, the *Aleinu*, in which Lurianic commentary references Rachel and Leah.

57. Gikatilla, *Gates of Light*, 303–4.

58. Gikatilla, *Gates of Light*, 305.

59. There are fascinating connections between this indeterminacy, and the one involving the two goats that are sacrificed at Yom Kippur, but this is beyond the limits of this chapter.

60. Acknowledgments: I have endless gratitude to the following people.

My teachers, who have taught me to read and travel through the texts of Kabbalah and chassidut: Rabbis Zalman Schachter-Shalomi z"l, Miles Krassen, Elliot Ginsburg, Marcia Prager and Itzchak Marmorstein.

My first teacher of Kabbalah, Professor Robert Haralick, who taught me to look at Kabbalah with an eye to pattern recognition, and to engage with it for ethical improvement and social justice.

Nathaniel Berman, who has called me to the Zohar over and over.

Karen Barad, who taught me so much about gender studies, queer theory, and quantum physics; shared endless conversations about Kabbalah, Jewish text, and physics, along with their expertise on diffractively reading quantum physics and Kabbalah through one another; and gave feedback on earlier drafts of this chapter.

My children, Mikaela Barad and Elisha Arafel, who have put up with my weirdness in so many ways.

Chapter Eleven

Toward a Bright and Messy Future

The Global Ecological Crisis, the Problem of Heteronormative Bias, and the Necessity of a Queer Ecological Imagination

Alex Carr Johnson

A few years ago, my husband and I signed our legal marriage papers inside of a giant sequoia tree. The hundred-and-fifty-foot-tall trunk had been hollowed out by fire long ago. The tree was still very much alive, indeed, fruiting fresh cones that season. There was something poetic, obviously so, about the gesture to make a lifelong commitment inside of a living thing that would likely live a hundred times as long as we would. It would live through many thousands of seasonal cycles in a similar fashion as we experience our days.

My parents were there to witness the ceremony. In fact, my mother was the officiant, the gesture holding obvious import, so too, the decision to carry out the ceremony on my husband's parents' wedding anniversary. We wished to honor both sets of our parents, the people who had partnered together, made us, raised us, and even now still supported us in love and guidance as we navigated our lives. *We humans pass on so much through the generations*, we were saying with our marriage, *knowledge, culture, commitment, beliefs, identity, and genes*.

Oh, wait. Let's talk about the genes part. You see, we were two men marrying each other, our wedding a queer one. The most obvious difference between our parents' marriage and ours? They could make babies together, as had their parents, and their parents before that, *ad origin*. We could not. It would be impossible for my chosen life partner and me to maintain the long line of my genetic progenitors into the next generation.

My husband and I were park rangers at the time, the ceremony held inside the boundaries of Yosemite National Park, one of the first places on the planet set aside for the protection of a living landscape. The other attendees at our wedding were other park employees, all of us dedicated to protecting and preserving the hollow sequoia in which we were marrying, along with the entire grove surrounding it, the incense cedars and juniper beyond the grove, the granite domes beneath them, and every other more-than-human thing that made up the landscape all the way across the Sierra Nevada mountains to the Great Basin desert far to the east.

As it happens, my husband and I both work in the field of conservation, our lives and livelihoods devoted to preserving and protecting the natural world in perpetuity. We spend much of our days engaging with a future world of which we as individuals will not be a part. At the time we married, we spent our days restoring Yosemite's high alpine meadows from the impact of humans, returning conditions so that the living things, the trees, meadow vegetation, and wildlife, could carry on for generations to come, and so that future generations of humans could come to recreate and enjoy the wildness as people had for generations before them. Yet, paradoxically, those future generations for which we worked would not include our own biological children.

THE PROBLEM OF HETERONORMATIVE BIAS IN CONSERVATION BIOLOGY

"Future generations" is a loaded term for queer people, as it often represents explicit assumptions regarding the alignment of an individual's reproduction success and that individual's worth. We are each bags of flesh and bone, the argument goes, only useful for our ability to pass along our best-fit genes from past to future.[1] This concept is called "fitness," and any organism's fitness has often been understood synonymously with the value of that organism evolutionarily. Of course, most people, including most scientists, do not claim that an individual's Darwinian fitness equates one-to-one with that individual's worth. However, over recent history, there have been remarkable and dangerous exceptions. The most obvious example is the phenomenon of eugenics, which sought to apply the theories of Darwinian evolution to the development of human societies. Nearly every world power during the first half of the 20th century carried out some form of eugenics programs which collectively harmed or killed hundreds of thousands of people across the globe.[2] Women, minorities, and queer people were disproportionately affected by these programs.

While eugenics is an extreme example of Darwin-gone-wrong, it is one worth bringing to bear as we assess the science underpinning one of humanity's greatest challenges over the next century: to protect and preserve biodiversity even as human activity is quickly expediting the planet's sixth mass extinction. A fundamental challenge of the field of modern conservation biology is that it operates under the starkest of calculi: which individuals, which populations, and which locations are of greatest value, and therefore, will be privileged with the greatest degree of protection? In short, it is an exercise of triage.[3]

Here's where the problem of heteronormative bias in conservation arises: if the individual organisms of greatest value are those with the greatest fitness, then those that appear to be less "fit," like those who exhibit homosexual behavior or other gender non-conformity, are inherently less valuable. Consequently, conservation biologists risk falling into the same ethically malignant pitfalls as the eugenists have in the past, arguing that queer individuals, human and non-human-alike, lack value, or in fact, pose risks to a given population's biological well-being.[4] The negative consequences of this sort of thinking are many, both scientifically and culturally, and which I will further explain over the course of this chapter.

A few of the most pertinent for the scientific field of conservation biology are the scientific justification for prejudice of non-heteronormative organisms; a correlated prejudice of non-heteronormative humans; and the inhibition of scientists to investigate more complex and nuanced understandings of biological fitness which may better explain the existence and evolutionary role of non-heteronormative behavior in species where it has been observed.

THE PROBLEM OF HETERONORMATIVE BIAS IN SOCIAL-CONSTRUCTS OF "FUTURE GENERATIONS"

For those of us living in a post-marriage-equality society, it's easy, and indeed convenient, to project legal equality to mean universal equality. *Now that gay people can marry, they are just like us*. This belief, however, would be a mistake: a classic colorblind fallacy.[5] There are and will always be material and embodied differences between heteronormative embodiment and queer embodiment worth acknowledging, honoring, and understanding, with reproduction one of the more immediate examples.

It is also convenient to laugh at arguments oppositional to marriage equality, particularly conservative claims that "gay" marriage as an institution is invalid due to the inability for gay couples to reproduce and raise biological children:[6] since surrogacy, adoption, and any variety of non-traditional family structures can provide the opportunity for gay people to reproduce and raise children. The existence of non-reproductive heterosexual married couples offers another easy refutation. In short, all coupled partners, including queer ones, currently have the same freedoms to create family structures of their choosing, many of which can look and feel very similar to heteronormative family structures.

Even so, I believe biological reproduction, as both a physical phenomenon and socially constructed concept, impacts the lives of queer people in ways that have yet to be properly critically examined. We must occupy a queerer middle space from which to examine the social and psychological implications of queer people's identities, embodiments, and partnerships which tend to form independently of the need to co-mingle genes to produce a biological child. Specifically, how does this basic circumstance of many queer partnerships affect how queer people individually and collectively relate to the future of the living world in which we live? Further, how might a queer examination of the socially constructed concept of generational reproduction prove to be a helpful tool for larger social concepts of our shared future, individual roles in proactively shaping that future, and the ethical reasons for taking the personal agency to do so?

In overestimating the universality of marriage and all forms of long-term partnership more generally, we also risk overlooking the unique lived experience of queer people in regard to their relationship to the future and their roles in shaping and caring for a future world. My marriage to a man inside of a sequoia was not an elaborate re-creation or simulacrum of traditional marriage between people of opposite genders. Instead, queer partnerships contain distinct tensions between their embodiment and the social mores and biases projected upon them. The tensions that most interest me now, as I remember that moment we signed the papers inside of the giant tree, have to do with how each of the participants and attendees viewed the future, how the act of publicly affirming our love for another man affected my husband and role in shaping that future, and how our queer identities and relations could affect the reasons why we would care about that future in the first place.

In recent history, with the rise of the modern environmental and conservation movements, practitioners continue to use the concept of "future generations" ubiquitously as a way of moving people to care, and correspondingly act, to protect the environment on behalf of their children and grandchildren. Simultaneously, conservation biologists have proven particularly adept at extending heteronormative views of sexuality, gender, and mating to their observations of human and other species' behavior.[7]

But what about queer people like my husband and me, who are unable to reproduce together biologically and less likely to produce biological children of our own? If, in fact, people are motivated to care about the future of the world primarily because they care first and foremost about themselves and their own biological children, then wouldn't it be the case that queer people wouldn't or even shouldn't care about the ecological future of the planet, not to mention the future generally?

This deductive reasoning is patently invalid based on the lived experience of many queer people who care deeply for the ecological health of the more-than-human world, both at present and far into the future. This deeply rooted compassion for a future world arose within us and continues to exist independent of our goals, hopes, and prerogatives regarding producing and raising children.

My husband's and my altruism, as exhibited by our evidenced environmental compassion, drives biologists nuts. Generally speaking, the evolutionary advantage of altruism, why individuals dedicate so much of their own resources to other individuals without any apparent individual gain biologically, has proven to be a rich field of scientific inquiry, pushing back on mechanical-flavored-orthodoxy of evolution as a process of each individual left to their own devices, good looks, and/or guile.[8] *If life is ruled by "survival of the fittest,"* the evolutionary biologists who study altruism ask, *then why would individuals sacrifice their lives for strangers?* Presiding theories on altruism, however, ultimately reinforce the less-obvious genetic advantages to taking care of others who are not you, your mate, or your children. Research suggests people are still more likely to exhibit altruistic behavior for their closest kin, siblings, parents, nieces, and nephews.[9]

In this context, E. O. Wilson, one of the most influential contemporary evolutionary and conservation biologists, has conceptually extended the biological advantages of familial altruism as an explanation for not only altruistic behavior itself, but also, interestingly, for other behaviors which do not at-face-value appear to be of evolutionary advantage for the individual. Case in point: homosexual behavior. According to Wilson, families with nonreproductive members have more aunts, uncles, and/or cousins to raise children, and correspondingly the family will be more successful in safeguarding its genetic heritage for future generations.[10]

This evolutionary rationale for queer behavior is helpful in providing reasonable, scientifically justifiable answers to some of the basic questions regarding my husband and my existence as queer people, and why we have the capacity to dedicate so much of our individual productivity to one another without apparent biological gain. These theories also provide a basic starting point for answering the larger question of concern here: my husband and I care about the future of our living world because, first, we have evolved to care about our family and our collective continued existence, and second, there is, in fact, high likelihood we will have kin, nieces, and nephews, and cousins, who will depend on an ecologically stable planet at some future date long after we have passed away.

Alas, forwarding a likely theory for biological causation of a given behavior still leaves us woefully far from understanding the socially constructed interpretations, treatment, and implications of said behavior. In regard to queer people, for instance, we have been and continue to be treated as uninterested and unrelated parties in determining healthy, sustainable, and ecologically healthy futures.

It is no secret that conservationists for the last several centuries have often failed at welcoming diverse people and perspectives into the ongoing discussion of how to protect and maintain the ecological health of our landscapes, watersheds, and globe.[11] In the last few decades, conservation NGOs, to their credit, have begun to acknowledge the paternal and colonial hegemonies within their legacies, and have critically assessed how they can better represent and include diverse communities, ethnicities, socio-economic background, and geographies within their membership and staff[12] and have begun carrying out policies and programs to better engage people that are not rich, white, and men.

Additionally, the field of environmental justice has proven to be an important tool for combating historic injustices, both of 1) the industrial forces that have disproportionately saddled people of less political power and social capital, very often people of color,[13] with the public and environmental health impacts of development, and 2) the blindness of the inequity from efforts from conservationists and environmentalists to mitigate impacts from said development.

The ways in which the fields of conservation, both the science and corresponding advocacy, have not only failed to welcome but have disenfranchised queer people, however, has not faced serious critical examination until very recently. As previously mentioned earlier in this chapter, conservationists have relied heavily on "future generation" language regarding reproduction as the primary reason for caring about the environment. They have also extended biological mechanisms as existential maxims. This is not the end of the disenfranchisement, however.

In her book *Strange Natures*, Nicole Seymour catalogues several other examples, including EarthFirst! leaders celebrating the AIDS epidemic as a method for controlling global human populations.[14] This type of argument undermines the ethical standing of the movement as a whole, and can very quickly become a slippery slope toward immoral social programs like sterilization and dangerous concepts verging on eugenics. This type of risk in valuing individual worth based on their reproductive abilities is at the heart of conservation biology's ethical failures and pitfalls of the past. The scientific record is rich in examples of biologists extending their own biases of homosexual behavior as not only an evolutionary disadvantage, but morally and ethically abhorrent.[15] These micro-aggressions slowly accumulate to produce a body of work that excludes other, queerer, interpretations for biological behavior, while also disenfranchising people who may be in a better position to contribute queer additions and corrections to the science. A queer biologist, for instance, may be more likely to observe non-heteronormative behavior and physical embodiment in the organisms they are observing, which has proven to be a challenge to biologists in the past.[16] A queer biologist may also be more likely to consider other evolutionary values that improve the Darwinian fitness of an individual organism, a particularly relevant line of inquiry as biologists continue to confirm the validity of other more nuanced mechanisms of evolution, including group fitness and sexual and social selection.[17]

Compounding these factors, public queer expression has, in recent history, largely been restricted to urban centers, producing all sorts of geographic and cultural positive feedback loops that further separated queer people, communities, and cultures, from access to the rural, largely still ecologically intact, and wild landscapes where much of the research within the field of conservation was taking place.

NEGATIVE IMPLICATIONS OF HETERONORMATIVE BIAS IN CONSERVATION

It's worth noting that contemporary discourse on queer theory is taking place at a very specific time on a very specific planet. Over the last several decades, thanks to the accomplishments of researchers within the many fields of the biological and earth sciences, we have achieved historic levels of understanding regarding the unprecedented and rapid ongoing changes occurring to biophysical systems around the globe. Climate change is but one example of massive global change reshaping the globe. Resource extraction, increasing populations, residential development and urbanization, deforestation, and climate change all fueled by growth-dependent socio-economic structures are just a few of the destructive and destabilizing forces affecting communities and landscapes around the planet.[18] With these changes come severe and immediate negative implications to the well-being of all human and non-human communities. The data support not only massive changes but also the urgent need for equally massive global political and social action to mitigate their impacts.[19] In many ways, the stakes of the modern conservation movement are life or death for individual organisms, species as a whole, and functional ecosystems for all life on Earth.[20]

Current extinction rates are 1000 to 10,000 times greater than the baseline rates through the history of life on Earth's history. This means that 30–50% of all species on Earth may be extinct within 50 years.[21] Each of these species has a unique niche within its constituent ecosystem, and the ecological effects of mass extinctions are impossible for anyone to know with certainty. Humans rely entirely on the biophysical systems which provide ecosystem services like clean and copious water, clean, oxygenated air, and protection from natural disasters like floods, hurricanes, droughts, and fires.[22]

This is not meant to be exaggeration for the sake of emotional appeal, but instead, I mean to put this conversation on the most honest footing regarding the state of biological function on this planet at this moment in time. Across the globe, biologists are seeing a precipitous decline in both population numbers and diversity. In fact, we are collectively living in the sixth mass extinction of the planet, with extinction rates seen rarely in the history of the planet.[23] We collectively possess an urgent need to understand how biophysical systems work on this planet, the causes for destabilization of those systems, and the best strategies for re-stabilization. What, then, in this urgent moment, are the more utilitarian purposes of queering conservation? Indeed, what necessary usefulness does queer theory provide to the field of conservation? Is there any usefulness at all?

A QUEERER ECOLOGY: MORE ACCURATE, MORE USEFUL, AND MORE INCLUSIVE

The direct and indirect consequences of challenging conservation's heteronormative biases include an increase in the field's capacity to: accurately assess and understand the extent of biodiversity; articulate the complexity of the challenges of preserving biodiversity; and inspire engagement in a more-inclusive conservation ethic which includes individuals and communities who are not moved by the heteronormative notions of protecting the planet for future generations.

Queer theory offers a reasoned critique and check for many of the most basic heteronormative biases throughout conservation biology's history. When a biologist sees a partnered pair of a non-dimorphous species (in which all adults appear identical to the human eye, regardless of sex), should they assume that it is an opposite-gendered pair? No.[24] When a biologist sees same-sex mounting in ungulates, should it be interpreted as aggression? No.[25] When a biologist observes an individual organism, should it assume the sex based on plumage, coloration, or the appearance of amorphous horns? No,[26] no,[27] and again, no.[28] In fact, for decades, biologists made all of these assumptions based on their own heteronormative biases. A queer conservation biologist, or one who has the capacity to observe and assess the living world through a queer perspective, is more likely to acknowledge a greater spectrum of behaviors, forms, and functions of the living organisms, behaviors, and relationships. The world is far queerer than we have supposed.[29] The consequences of extending queer theory to conservation, I believe, will produce a number of possible positive outcomes including a more critical examination of the various roles sexual behavior plays in biological systems and evolutionary processes, new insights into the mechanisms with which to conserve species, and more inclusive and universally compelling frameworks for communicating the global ecological crisis.

As I've described above, heteronormative biases have hindered biologists' abilities to observe non-heteronormative behavior and gender non-conformity in the organisms that they are observing. While the acute consequences of missing data points is troubling in the context of individual studies, I'm more concerned about the universal consequences of entire missing sets of data, whole segments of populations that are overlooked, marginalized, or devalued because they are non-heteronormative. I see no reason why this type of blind-spot within the science would not extend to the analysis of the data, and the conceptual framework in which the observations are interpreted. The foundational concepts of conservation biology are that 1) there exists genetic diversity within populations of all living things, 2) there exists a minimum number of individuals needed to maintain that genetic diversity, and 3) there exist mechanisms and tools within human capacity for promoting the continued existence of any given population above that minimum threshold. How is one to trust the science of conservation biology, which determines which organisms, populations, and locations are of greatest value to evolutionary well-being—if the science itself appears skewed by heteronormative bias? A queerer understanding of ecology would quite literally *allow for the existence* of the full spectrum of biological behaviors and embodiments of sexual capacity within living things, consequently offering more informed recommendations regarding policies for managing efforts to conserve biological diversity on this planet. After all, sexual diversity is a central piece of biological diversity.

Evolutionary biology as a science is a relatively young one, and biologists are continuing to refine their understanding of the complex and messy processes that shape the evolution of life. Queer-informed research of evolution could lead to new understandings of not only the effects of non-heteronormative behaviors and embodiments, but in fact the evolutionary values of these. Conservation strategies for protecting and preserving biodiversity would then include in them the necessary protection of those newly valued behaviors and the individuals that exhibit them. These queerer biological theories and conservation strategies would most certainly also inform social construction of the value and worth of non-heteronormative human individuals as well.

This last point is an essential one if we are to find long-term and sustainable conservation solutions. We as a global society making global decisions regarding the future of life on the planet need to understand not only the evolutionary rules and imperatives to keep species in existence, but we also need to understand how to communicate these needs in ways that honor the remarkable diversity of behaviors, cultures, identities, and communities contained within the current polyglot of humankind. In fact, conservationists not only have to figure out the right language to communicate these needs, they must also integrate into as many diverse communities as possible, cultivating conservationists regardless of whether they take on the title. To be sure, global conservation success will only come if people fully incorporate conservation practices and identities into their daily lives *while also* supporting state, national, and international policies that promote preservation of biodiversity conservation. Our only hope for this sort of state shift to occur is if we radically re-envision the basic value systems with which humans orient ourselves to each other, the more-than-human world, and the shared future of all life on this planet.

As a queer person and a conservation practitioner, I believe this radical re-envisioning must include a queerer ecology which will empower biologists and conservation practitioners to speak about the values of biodiversity in terms that are inclusive to people along the full sexual and gender spectrums. Queer people have historically been marginalized, underestimated, and overlooked, just like our non-human corollaries, and this marginalization can pose an unbridgeable divide. Even if queer people do care about the more-than-human living world, why would we support conservation science and policy that inherently devalues our own lived existence? Conservationists who were practicing a queerer conservation biology which recognized the possible values of non-heteronormative behavior and embodiments would have the platform from which to speak in inclusive ways that could inspire the full spectrum of human populations. This is the conservation that could create the state shift necessary to stem the massive ongoing loss of global biodiversity.

THE UNSETTLED FIELD OF CONSERVATION BIOLOGY

After getting married, my husband and I moved to a small mountain town in Colorado, where I took a job with the local conservation group. By complete chance, we moved in next door to Dr. Michael Soule, often referred to as the father of conservation biology.[30] He in fact coined the name for the field.[31] Some of his other accomplishments include developing the initial genetic science to apply evolutionary principles to specific populations[32] and later, the idea of landscape-scale connectivity.[33] We met early on, after his wife attempted to construct a greenhouse in front of our only south-facing window. We very quickly went outside to introduce ourselves, and fortunately, very quickly became close friends. Shortly thereafter, Michael and I began meeting to talk about evolution, conservation, and the future of the planet over tea.

I was not blind to the inherent tension of being a queer conservationist talking about evolution with a luminary in the field of conservation biology and genetics. After all, Michael developed the field which I've been critiquing for the last six pages. He was dear friends with radical conservationists like Dave Foreman who founded EarthFirst!, and who once applauded the AIDS epidemic for its positive effects for conservation.[34] Michael also happened to fit the basic characteristics of the maligned hegemonic conservationists: he was white, male, straight, and apparently in every way a biologically fit individual, having had four children himself, each with children of their own. During one of our afternoon tea meetings, I asked him where would be the safest place for his grandchildren to live in the face of climate change and the ongoing global ecological crisis. The question, upon asking, struck me as particularly one-directional: interesting for me, satisfying a curiosity, but not particularly useful or impactful to my own life, plans, or intentions.

I proceeded to ask Michael many more questions over the years between then and now, and the conversation became more back and forth. We spoke about the actual conceptual mechanics of the field, the differences between sexual selection and natural selection, for instance, or the boons and risks of incentivizing conservation through economic means. As you might expect, my questions led most often toward additional questions rather than answers. My clearest conclusions from our friendship has been that the field of conservation biology, the one Michael helped found, is a deeply unsettled one, with a number of different factions of scientists passionately disputing and debating not only the science, but even more, the application of that science in the real world in real time.

As much as any modern field of science, conservation biology is an unsettled one. Consensus within the field grows exceedingly more clear each day: biological diversity is in decline due to massive global anthropogenic impacts. The implications of these data are also quite clear: continued specific and universal destabilization of biophysical systems, rapid loss of species diversity, and continued decline in ecological composition and function. The scientific evidence describes a bleak future for life on this planet, let alone for the majority of humankind. Yet the technical, social, political, and ethical responses to this growing body of evidence are proving to be exceedingly muddled. It's a true modern dilemma of the highest order: we know the problems, we know the solutions, but we have no idea how to move forward with any unified response while holding on to our collective humanity.

The dilemma that has risen out of the field of conservation biology is central to the modern dilemma of being human on this earth in this moment. E. O. Wilson, the prolific biologist that he has been, has also forwarded a theory on the capacity for humans to love the more-than-human living world. He calls it biophilia, and if he is correct, then humans, like all living things, possess a biophilia at the very base of our being.[35] We love our home. We love our family. And yet, so long as we continue to live our modern lives in our modern societies, which even with the most radical "lifestyle choices" of becoming a vegetarian, locavore, ditching your car, it is all but a certainty that we are facilitating the death of the *bio*. We are living in a biodiversity crisis seen few times on this planet. By our very existence, we are perpetuating the demise of the thing we hold most dear.

Thanks to conservation biologists and conservationists, like Michael Soule and E. O. Wilson, we do largely know the cause of the problems as well as the ecological solutions to the biodiversity crisis. Stop habitat fragmentation. Decrease consumption and human population increase. Manage landscapes for connectivity. Yet many of those solutions, in practice, have proven to be infeasible, due to a complex milieu of political, social, and moral inertia. It's the *how* that we can't get right. And the *why*.

After all of those conversations with Michael, afternoons looking out at the mountains, or out floating down one of our rivers close to our shared home, I've really only come to two clear conclusions: 1) the father of modern conservation biology largely believes we're failing at protecting and maintaining the ecological health of the planet, and 2) we still need new, radical, and creative ways of re-imagining our relationships to each other, to the more-than-human world, and to the future of our planet. In fact, we need mothers of conservation, brothers and sisters, and aunts and uncles, and queens and kings of conservation, maybe even, probably, in drag.

ALTERNATIVE RESPONSES TO THE BIODIVERSITY CRISIS

I and my fellow "queer ecologists" are not the first to identify an urgent need for staging a more effective response in the face of the current ecological catastrophe, its existence supported by the science of conservation biology. A number of other philosophies and movements have sprung up to do just that. One school of thought has taken the title of "New Conservation," whose most public and ardent supporter has been Peter Kareiva, the former scientific head of The Nature Conservancy.[36] Much to Michael Soule and other traditional conservationists' chagrin,[37] Kareiva has refuted the need to elevate the intrinsic value to the central motivation for conservation when economic value can provide much more tangible reason to conserve, and motivates corporate entities who have the incentive to protect their assets and the financial power to do so. Kareiva, along with fellow adherents to this New Conservation, have formed a "Natural Capital" movement, which they believe overcomes many of the historic inertia and injustices associated with conservation.

Another conceptual challenger to make recent waves within conservation circles is the "Garden Earth" school, as recently popularized by Emma Marris in her book, *The Rambunctious Garden: Saving Nature in a Post-Wild World.*[38] The basic premise is clear: similar to the New Conservationists, the Earth Gardeners believe it is time to accept the fact that we are the planet's stewards, and if we take on the role in earnest, we can allow for human and ecological needs simultaneously, much as a permaculturist does on their own little plot of land.

A third response, a direct counterpoint to the first two, calls itself "The Dark Mountain Project,"[39] and takes its name from the Robinson Jeffers poem "Rearmament," whose final lines describe the modern human experiment as "the Disastrous rhythm, the heavy and mobile masses, the dance of the / Dream-led masses down the dark mountain."[40] The Dark Mountain Project is meant to be a direct reckoning with the current inevitability of the crisis itself, and rather than attempt to actively find a solution to it, the project seeks to build new ethical frameworks through which to maintain one's humanity in an inhumane time.

In a certain way, this movement appears the most emotionally honest and true to the scientific facts relative to the other alternatives and supplements to the field of conservation. Even so, The Dark Mountain Project, at times, repeats the same social injustices carried out by the field of conservation before it.[41] It takes a very specific privilege of identity, economics, and place, to be able to determine that there is no use in working toward finding sustainable solutions that create stability globally, providing stability to any number of people who are incapable of mitigating, escaping, or adapting to the effects of rapid global change.

As I've written elsewhere, I don't believe that queer ecology is a field in and of itself, so much as a call for the queering of ecological sciences, ethics, and actions.[42] As such, it largely circumvents the so-oft divisiveness between different intellectual camps. Queer ecology is not a set of beliefs, but rather a call for critically examining the physical locations and bodies from which we each interact with the world, and for radically re-evaluating the biases, histories, privileges, and realities associated with that embodiment.

A QUEER ECOLOGICAL IMAGINATION

Queer theory can improve the science of conservation by allowing for the existence and value of non-heteronormative behaviors and embodiments, but as is likely clear to the reader at this point, I believe that solutions to our global ecological crisis will also depend on the ethical and philosophical framework with which humans act on behalf of the future. So how, then, do we re-imagine a queerer relationship between ourselves, the more-than-human world, and the future?

Nicole Seymour offers what I have found to be the most promising idea in the form of what she calls queer ecological imagination. A queer ecological imagination would not be rooted in biological fitness at all, which itself demands a heteronormative futurity that values reproduction as the method for individuals to develop their own sense of meaning and care. Instead, a queer ecological imagination would root itself in a form of care that operates outside the realm of human-specific solutions, identities, or recompense. Seymour forwards the notion that queer theory, which eschews "stable identities, epistemologies, and ontologies . . . might lend itself most effectively to empathetic, politicized advocacy for the non-human natural world." A transgender person, Seymour continues, who has "often been treated as subhuman," may be more likely to empathize with the more-than-human living world and the other non-human creatures that live within it. This person may also be more likely to empathize with the other subjugated people within their society who may be most negatively affected by the global changes which we are now witnessing. As Seymour articulates:

> In the so-called real world, such a scenario would be the most delicious, and queerest, irony of all . . . that it might be, say, the childless or unmarried person, or the queer who "lives in the now," or the person whose body is constructed through "unnatural technology"—in short, the person who has everything to lose and nothing to gain—who cares most for the future well-being of non-human and human others.[43]

For me, the answer to that question has everything to do with a very particular sequoia tree. My husband and my commitments inside that tree had very little, almost nothing, to do with legal validation, and had mostly, nearly everything, to do with our commitment to a family of living things that extends much further than each other and our non-existent children, further than our parents and siblings and their children, further than the friends in witness, further even than the trees and the deer and the mountains. We were also committing to a future, of which we knew very little about, but which we would dedicate our lives to making richer, messier, more diverse, and more just for our extended living family whose bounds we committed to expanding.

In regard to conservation, a queer ecological imagination allows for all individuals to identify themselves as part of their own family of living things, constantly in flux, far more diverse than what our social and scientific institutions have acknowledged historically, and far more inclusive as well. How far might a person be able to extend their own identity as a family member, and their corresponding sense of care and compassion, once freed from the socially constructed expectations of familial responsibility and obligation? Could a person care for their neighbor's child as much as their own? What of that child's future children? What of those children's neighbors? What of the river whose water those neighbors will drink? What of the fish in that river? The otters and herons that feed on those fish? The mountains from which that river flows? The trees growing on the side of those mountains? One particular tree, long ago burned hollow from an ancient fire, still dropping fresh cones every spring?

In the *Power of Myth*, Joseph Campbell explains to Bill Moyers just this sort of outward expanding sense of compassion:

> "So when we say, 'Save the earth,' we're talking about saving ourselves," says Moyers.
>
> "Yes," says Campbell. "All this hope for something happening in society has to wait for something in the human psyche, a whole new way of experiencing a society. And the crucial question here, as I see it, is simply: With what society, what social group, do you identify yourself? Is it going to be with all the people of the planet, or is it going to be with your own particular in-group?"[44]

I would argue queer theory provides one of the best tools for breaking down the entirely constructed boundaries between the in-group and out-group, self and other, natural and unnatural, human and more-than-human. In the process of saving ourselves, we might not only save the world, but realize the two, we and the world, are in fact not separate in any way, ecological or otherwise.

CONCLUSION: TOWARD A BRIGHT AND MESSY FUTURE

Over the course of this chapter, I've attempted to circumscribe the shape of a great and urgent need, and one which I believe can only be successfully answered with the help of a critical queer approach. Not only can queer theory broaden the scientific boundaries in which conservation biologists are operating, by allowing for a more comprehensive understanding of biological diversity and the evolutionary processes which shape it, but it can also expand the audience of those receptive to the conservation strategies developed from the science.

Even more, queer theory can offer the ethical scaffolding from which to hang our understanding of how and why people care about the future of life on Earth. In fact, a queer approach to the biodiversity crisis—and our global ecological dilemmas more generally—could fundamentally shift our collective understanding of humans' role on this planet. Queer theory demands more uncertainty about the way in which consequences ripple out into complex systems like our living Earth. How might our observations, values, and actions affect other bodies in positive and negative ways? How might we reimagine our relationships to each other, the more-than-living world, and our shared presence in it, past, present, and future?

As Seymour concludes, "the kind of empathy that environmentalism at large calls for so urgently right now is by definition queer, even when not directly linked to (homo) sexuality or sexual issues: one must care for nameless, faceless future beings, including non-humans, to which one has no domestic, familial, or financial ties."[45]

May we heed the call, and may the future be bright, boisterous, queer, and messy.

WORKS CITED

Allen, Garland E. "The Social and Economic Origins of Genetic Determinism: A Case History of the American Eugenics Movement, 1900–1940 and Its Lessons for Today," *Genetica* 99 (1997): 77–88.

Bagemihl, Bruce. *Biological Exuberance: Animal Homosexuality and Natural Diversity*. New York: St. Martin's Press, 1999.

Barnosky, Anthony D., Nicholas Matzke, Susumu Tomiya, Guinevere O. U. Wogan, Brian Swartz, Tiago B. Quental, Charles Marshall, Jenny L. McGuire, Emily L. Lindsey, Kaitlin C. Maguire, Ben Mersey, and Elizabeth A. Ferrer, "Has the Earth's Sixth Mass Extinction Already Arrived?" *Nature* 471 (2011): 51–57, accessed April 12, 2017, doi:10.1038/nature09678.

Bottrill, Madeleine C., Liana N. Joseph, Josie Carwardine, Michael Bode, Carly Cook, Edward T. Game, Hedley Grantham, Salit Kark, Simon Linke, Eve McDonald-Madden, Robert L. Pressey, Susan Walker, Kerrie A. Wilson, Hugh P. Possingham, "Is Conservation Triage Just Smart Decision Making?," *Trends in Ecology & Evolution*, 23.12 (2008): 649–654, accessed July 29, 2017, doi: 10.1016/j.tree.2008.07.007.

Bullard, Robert D. *Dumping in Dixie*. Boulder, CO: Westview Press, 2008.

Campbell, Joseph, Bill Moyers, and Betty Sue Flowers. *The Power of Myth*. New York: Doubleday, 1988.

Center for Biological Diversity. "The Extinction Crisis," accessed July 29, 2017, http://www.biologicaldiversity.org/programs/biodiversity/elements_of_biodiversity/extinction_crisis/

Dawkins, Richard. *The Selfish Gene*. Oxford: Oxford University Press, 1976.

Gardner, A., S. A. West, and G. Wild, "The Genetical Theory of Kin Selection," *Journal of Evolutionary Biology*, 24: 1020–1043, accessed May 15, 2017, doi:10.1111/j.1420–9101.2011.02236.x.

Haldane, J.S. *Possible Worlds*. New York: Routledge Press, 2017.

Johnson, Alex. "How to Queer Ecology: One Goose at a Time." *The Environment in Anthropology: A Reader in Ecology, Culture, and Sustainable Living* (2016): 310.

Kingsnorth, Paul, and Dougald Hine. *Uncivilisation: The Dark Mountain Manifesto*, accessed July 28, 2017, http://dark-mountain.net/about/manifesto/

Kloor, Keith. "The Battle for the Soul of Conservation Science." *Issues in Science and Technology* 31, accessed July 20, 2017, http://issues.org/31–2/kloor/

Kurzban, Robert, Maxwell N. Burton-Chellew, and Stuart A. West. "The Evolution of Altruism in Humans." *Annual Review of Psychology* 66 (2015): 575–599.

Loreau, Michel, Alfred Oteng-Yeboah, M. T. K. Arroyo, D. Babin, R. Barbault, M. Donoghue, M. Gadgil, C. Häuser, C. Heip, A. Larigauderie, K. Ma, G. Mace, H. A. Mooney, C. Perrings, P. Raven, J. Sarukhan, P. Schei, R. J. Scholes, and R. T. Watson. "Diversity Without Representation," *Nature* 442 (2006): 245–246, accessed March 16, 2017, doi:10.1038/442245a.

Marris, Emma. *Rambunctious Garden: Saving Nature in a Post-Wild World*, New York, NY: Bloomsbury USA, 2011.

Mau, Bruce. *Massive Change*. New York: Phaidon, 2004.

Miller, B., M. E. Soulé, and J. Terborgh, "'New Conservation' or Surrender to Development?" *Animal Conservation* (2014).

National Wildlife Federation. "Ecosystem services," accessed July 28, 2017, https://www.nwf.org/Wildlife/Wildlife-Conservation/Ecosystem-Services.aspx

Peterson, Jodi, "Parks for All?" *High Country News*, May 12, 2014, accessed July 10, 2017, http://www.hcn.org/issues/46.8/parks-for-all.

Seymour, Nicole. *Stranger Natures: Future, Empathy, and the Queer Ecological Imagination*. Champaign: University of Illinois Press, 2011.

Smart, Carol. *The Ties That Bind: Law, Marriage and the Reproduction of Patriarchal Relations*. New York: Routledge, 2013.

Smith, Daniel. "It's the End of the World as We Know It . . . and He Feels Fine," *New York Times*, April 17, 2014, accessed June 5, 2017, www.nytimes.com/2014/04/20/magazine/its-the-end-of-the-world-as-we-know-it-and-he-feels-fine.html.

Soule, Michael, *The Collected Papers of Michael E. Soulé: Early Years in Modern Conservation Biology*. Washington D.C.: Island Press, 2014.

Stubblefield, Anna. "'Beyond the Pale': Tainted Whiteness, Cognitive Disability, and Eugenic Sterilization." *Hypatia* 22, no. 2 (2007): 162–181.

Williams, Monnica T. "Colorblind Fallacy Is a Form of Racism." *Psychology Today*, accessed March 28, 2017, https://www.psychologytoday.com/blog/culturally-speaking/201112/colorblind-ideology-is-form-racism

Wilson, E. O. *Biophilia*. Cambridge: Harvard University Press, 1984.
———. *Sociobiology: The New Synthesis*. Cambridge: Harvard University Press, 2000.

NOTES

1. Richard Dawkins, *The Selfish Gene* (Oxford: Oxford University Press, 1976).

2. Garland E. Allen, "The Social and Economic Origins of Genetic Determinism: A Case History of the American Eugenics Movement, 1900–1940 and Its Lessons for Today," *Genetica*, 99 (March 1997), 77.

3. Madeleine C. Bottrill, Liana N. Joseph, Josie Carwardine, Michael Bode, Carly Cook, Edward T. Game, Hedley Grantham, Salit Kark, Simon Linke, Eve McDonald-Madden, Robert L. Pressey, Susan Walker, Kerrie A. Wilson, and Hugh P. Possingham, "Is Conservation Triage Just Smart Decision Making?," *Trends in Ecology & Evolution*, 23.12 (2008): 649, accessed July 29, 2017, doi: 10.1016/j.tree.2008.07.007.

4. Anna Stubblefied, "'Beyond the Pale': Tainted Whiteness, Cognitive Disability, and Eugenic Sterilization." *Hypatia* 22, 2 (2007): 162–181.

5. Monnica T. Williams, "Colorblind Fallacy Is a Form of Racism," *Psychology Today*, accessed March 28, 2017, https://www.psychologytoday.com/blog/culturally-speaking/201112/colorblind-ideology-is-form-racism

6. Carol Smart, *The Ties That Bind (Routledge Revivals): Law, Marriage and the Reproduction of Patriarchal Relations* (New York: Routledge, 2013).

7. Alex Carr Johnson, "How to Queer Ecology* One Goose at a Time." *The Environment in Anthropology: A Reader in Ecology, Culture, and Sustainable Living* (2016): 310.

8. Robert Kurzban, Maxwell N. Burton-Chellew, and Stuart A. West, "The Evolution of Altruism in Humans," *Annual Review of Psychology* 66 (2015): 580.

9. A. Gardner, S. A. West, and G. Wild, "The Genetical Theory of Kin Selection," *Journal of Evolutionary Biology*, 24: 1020, accessed May 15, 2017, doi:10.1111/j.1420–9101.2011.02236.x.

10. E. O. Wilson, *Sociobiology: The New Synthesis* (Cambridge: Harvard University Press, 2000).

11. Michel Loreau, Alfred Oteng-Yeboah, M. T. K. Arroyo, D. Babin, R. Barbault, M. Donoghue, M. Gadgil, C. Häuser, C. Heip, A. Larigauderie, K. Ma, G. Mace, H. A. Mooney, C. Perrings, P. Raven, J. Sarukhan, P. Schei, R. J. Scholes, and R. T. Watson, "Diversity Without Representation," *Nature* 442 (20 July 2006): 245, accessed March 16, 2017, doi:10.1038/442245a.

12. Jodi, Peterson, "Parks for All?" *High Country News*, May 12, 2014, accessed July 10, 2017, http://www.hcn.org/issues/46.8/parks-for-all

13. Robert D. Bullard, *Dumping in Dixie* (Boulder, CO: Westview Press: 2008).

14. Nicole Seymour, *Stranger Natures: Future, Empathy, and the Queer Ecological Imagination* (Champaign: University of Illinois Press: 2011).

15. Johnson, "How to Queer Ecology," 310.

16. Bruce Bagemihl, *Biological Exuberance: Animal Homosexuality and Natural Diversity* (New York : St. Martin's Press, 1999).

17. Wilson, *Sociobiology.*

18. Bruce Mau, *Massive Change* (New York: Phaidon, 2004).

19. Mau, *Massive Change.*

20. Anthony D. Barnosky, Nicholas Matzke, Susumu Tomiya, Guinevere O. U. Wogan, Brian Swartz, Tiago B. Quental, Charles Marshall, Jenny L. McGuire, Emily L. Lindsey, Kaitlin C. Maguire, Ben Mersey, and Elizabeth A. Ferrer, "Has the Earth's Sixth Mass Extinction Already Arrived?" *Nature* 471 (2011): 51–57, accessed April 12, 2017, doi:10.1038/nature09678.

21. "The Extinction Crisis," Center for Biological Diversity, accessed July 29, 2017, http://www.biologicaldiversity.org/programs/biodiversity/elements_of_biodiversity/extinction_crisis/.

22. "Ecosystem services," National Wildlife Federation, accessed July 28, 2017, https://www.nwf.org/Wildlife/Wildlife-Conservation/Ecosystem-Services.aspx.

23. "The Extinction Crisis."

24. Bruce Bagemihl, *Biological Exuberance: Animal Homosexuality and Natural Diversity* (New York : St. Martin's Press, 1999).

25. Bagemihl, *Biological Exuberance.*

26. Bagemihl, *Biological Exuberance.*

27. Bagemihl, *Biological Exuberance.*

28. Bagemihl, *Biological Exuberance.*

29. J. S. Haldane, *Possible Worlds* (New York: Routledge Press, 2017).

30. Keith Kloor, "The Battle for the Soul of Conservation Science," *Issues in Science and Technology* 31, 2 (2015), accessed July 20, 2017, http://issues.org/31–2/kloor/.

31. Michael Soule, *The Collected Papers of Michael E. Soulé: Early Years in Modern Conservation Biology* (Washington, D.C.: Island Press, 2014).

32. Soule, *The Collected Papers of Michael E. Soule.*

33. Soule, *The Collected Papers of Michael E. Soule.*

34. Seymour, *Strange Natures.*

35. E. O. Wilson, *Biophilia* (Cambridge: Harvard University Press, 1984).

36. Peter Kareiva and Michelle Marvier, "What is Conservation Science?" *BioScience* 62, 11 (2012).

37. B. Miller, M. E. Soulé, and J. Terborgh, "'New Conservation' or Surrender to Development?" *Animal Conservation* (2014).

38. Emma Marris, *Rambunctious Garden: Saving Nature in a Post-Wild World* (New York, NY: Bloomsbury USA, 2011).

39. Paul Kingsnorth and Dougald Hine. *Uncivilisation: The Dark Mountain Manifesto*, accessed July 28, 2017, http://dark-mountain.net/about/manifesto/.

40. Kingsnorth, *Uncivilisation.*

41. Daniel Smith, "It's the End of the World as We Know It . . . and He Feels Fine," *New York Times*, April 17, 2014, accessed June 5, 2017, www.nytimes.com/2014/04/20/magazine/its-the-end-of-the-world-as-we-know-it-and-he-feels-fine.html.

42. Johnson, "How to Queer Ecology," 310.
43. Seymour, *Strange Natures*, 184.
44. Joseph Campbell, Bill Moyers, and Betty Sue Flowers, *The Power of Myth* (New York: Doubleday, 1988): 182.
45. Seymour, *Strange Natures*, 185.

Chapter Twelve

Queering the Library of Congress

Carlos R. Fernández

For 5,000 years, temple priests, kings, philosophers, monks, and librarians have had the political and societal power to identify, select, and acquire materials for their collections, as well as the power to create various cataloging and classification systems to preserve this knowledge. Modern classification systems, such as the Library of Congress Classification System, help organize collections and make the retrieval of information more efficient for library patrons. However, these systems also unwillingly contribute to how patrons and society as a whole perceive the world. Since classification systems reflect a vision of the world as seen by those with the power to create categories, it is important to question how these systems influence societal norms on various subjects. This chapter attempts to apply the concepts used in queer theory to explore library science's use of library classification systems; and examines how queer theory can improve these systems.

The term "queer" is used as an adjective, a noun, or a verb. It is commonly used as "an adjective, meaning 'not normal,' or, more specifically, not heterosexual."[1] As an adjective, "queer" describes those whose sexuality is outside of society's heterosexual norm—those who are gay, lesbian, bisexual, and transgender.[2] As a verb, the term means the deconstruction of conventional understandings of sexuality and society. Furthermore, "In academic circles, to queer something is to analyze a situation or a text to determine the relationship between sexuality, power, gender, and the conceptions of normal and deviant, insider and outsider."[3] Queer theory draws from the school of philosophy and literary criticism known as deconstruction. Deconstruction "is a social analysis of who, why, and what produced a text; and analysis of what is said—and unsaid—through the language, form, structure, and style of a text."[4] Moreover, queer theory "challenges [the] basic tropes used to organize our society and our language."[5] Queer theory can be useful in addressing problems found in the Library of Congress Classification System.

Unfortunately, antiquated cataloging rules, based on historically deep-rooted societal views, reinforce racist and homophobic beliefs. For instance, some libraries still catalog books on homosexuality and transsexualism under medical and psychological disorders and deviancies. This chapter will attempt to demonstrate how updating, or queering, the Library of Congress Classification System can help make materials on sensitive topics more accessible to patrons. However, before one can discuss modern librarianship and classification systems, one needs to understand the history and uses of library classification systems.

THE ANCIENT NEAR EAST

Mesopotamia

Historians estimate that libraries first appeared around 3000 BCE in what is known today as the Fertile Crescent, which stretches from Mesopotamia's Tigris and Euphrates rivers all the way to the Nile River in Egypt. The oldest surviving records from this region were written roughly 5,000 years ago by the Sumerians on clay tablets, in an alphabet known as cuneiform. These cuneiform records were economic in nature. "Temple bureaucrats" had the responsibility of inscribing these records, which consisted of tallying transactions of crops, animals, and manufactured goods, as well as preserving religious hymns, prayers, and incantations.[6] According to Lerner:

> The temples did not discard their old business records, but deposited them in storerooms. These were intended for preservation rather than for frequent consultation; often the storerooms had no doorways and could be entered only by descending a ladder. Within these rooms, tablets were kept in wooden boxes, reed baskets, or brick receptacles. Attached to each container was a clay tablet serving as a label, listing the contents of the tablets within each container and the dates they covered. Like modern [libraries], the temples often generated more records than they had room for.[7]

These tablet collections continued to be preserved after the Sumerian culture was replaced by succeeding civilizations, first the Akkadians, and then the Assyrians, Persians, and Macedonians. In addition to preserving the Sumerian texts, Akkadian scribes created records and "lists of signs and bilingual glossaries" for future generations of scholars.[8] These lists, or catalogs, were also inscribed on clay tablets, and they listed the first line or the first few words of each tablet of a grouping of tablets in the collection. Unlike books today, these texts did not have titles. As mentioned above, each tablet was known by the first line of its text, similar to the contemporary practice of identifying an untitled poem in an anthology.[9] This method of identification is called incipit, and has led some scholars to mistake catalogs of tablets for ancient poetry. The following incipit catalog highlights the importance of understanding classification systems, since it was easily mistaken for an unintelligible Sumerian poem:

> Honoured and noble warrior
> Where are the sheep
> Where are the wild oxen
> And with you I did not
> In our city
> In former days
> Lord of the observance of heavenly laws
> Residence of my God
> Gibil, Gibil [the fire-god]
> On the 30th day, the day when sleeps
> God An, great ruler [sky-god]
> A righteous woman, who heavenly laws and commandments
> The King whom you bore . . .[10]

Furthermore, each tablet contained a colophon. Per the Oxford English Dictionary, colophon is defined as "The inscription or device, sometimes pictorial or emblematic, formerly placed at the end of a book or manuscript, and containing the title, the scribe's or printer's name, date and place of printing, etc."[11] In the case of ancient tablets, the "Colophon begins with the number of the tablet it is on. . . . The colophon often includes . . . the name of the local scribe who copied the text."[12] Following are some examples of colophons found in Hattusas, provided by Lionel Casson in his work, *Libraries in the Ancient World*:

- Eighth tablet of the Dupaduparsa Festival, words of the Silalluhi and Kuwatalla, the temple-priestess. Written by the hand of Lu, son of Nugissar, in the presence of Anuwanza, the overseer.
- Third tablet of Kuwatalla, the temple-priestess. Not the end. "When I treat a man according to the great ritual" [i.e., the first line used as title].
- Second tablet of Tudhaliyas, the great king, on the oath. End. This tablet was damaged; in the presences of Mahhuzi and Halwalu, I, Duda restored it.[13]

Like the Akkadians, the people of Ebla also preserved Sumerian tablets and created catalogs. In the remains of the ancient city of Ebla, located in northern Syria, 15,000 tablets were found in the ruins of the royal palace's storage room, which was the palace's equivalent to a library. The majority of these tablets contained commercial, legal, and administrative records, and served as a reference collection for the palace staff. However, there were also "dictionaries and syllabaries" which were used by scribes when editing literary texts, as well as "epic narratives, myths, hymns, incantations, rituals, and proverbs . . . manuals of botany, zoology, mineralogy, and mathematics."[14] Instead of using an incipit system of cataloging, the Eblaites used descriptors such as place names, the names of objects, grammatical texts, and bilingual dictionaries. In addition to being a repository for the preservation of Ebla's collected knowledge, the palace library served as a school for scribes, and as a place for the propagation of knowledge.

One of the first known librarians was King Esarhaddon of Assyria (681 to 669 BCE).[15] King Esarhaddon started his library out of his interests in magic and astronomy; he collected texts on rituals, prophecies, and omens.[16] His son, King Assurbanipal (688–627 BCE), continued his work, and became a master scribe, often boasting that he achieved "the highest level in the scribal art."[17] Furthermore, he also "collated, revised, and edited the texts" he requested from collections in other temples and kingdoms.[18] On the tablet containing the text of the Epic of Gilgamesh, Assurbanipal wrote the following in the colophon:

> Palace of Ashurbanipal, King of the World, King of Assyria, who trusts in Ashur and Ninlil, whom Nabu and Tashmetu gave wide-open ears and who was given profound insight. . . . The wisdom of Nabu, the signs of writing, as many as have been devised, I wrote on tablets, I arranged [the tablets] in series, I collated [them], and for my royal contemplation and recital I placed them in my palace.[19]

Subsequent ancient librarians had titles such as "Masters of the Books" or "Keepers of the Tablets," and they were the scribes and priests in ancient temples.[20]

The Egyptians, like their Mesopotamian contemporaries to the east, preserved their knowledge in libraries. However, the Egyptians mainly used papyrus as the medium for their writings. Though their papyrus scrolls have deteriorated over the millennia, much of their other writings were preserved on the walls of their temples and tombs, as well as on pottery shards. Egyptian libraries were located in temples, and it is believed that their collections served mainly theological and liturgical purposes.[21]

Library of Alexandria

In the third century BCE, the Ptolemy rulers of Egypt established the Library of Alexandria. Like its predecessors in Mesopotamia, the library was located in a religious institution, the *Mouseion*, which served as a shrine to the Greek Muses. However, this library was one of the first libraries to serve the public, though only those with the proper scholarly credentials. The library's "aim was to collect the *sum total* of Greek *literature*—the recorded expressions of *intellectual* activity."[22] As a result, it housed every Greek text known to the Western world, essentially becoming a repository for Greek culture.

Callimachus was the third librarian to take leadership of the library, between 260 and 240 BCE. During his tenure, he created one of the first library catalogs and named it the *Tables of Those Who Have Distinguished Themselves in Every Form of Culture and of What They Wrote*. Scholars call this work Pinakes for short. The Pinakes takes its name from the Greek word for tables or tablets, indicating that it was a compilation of the tablets used to label shelves of papyrus scrolls.[23] Callimachus's catalog covered the whole scope of the library's collection, which, in turn, was composed of the whole of Hellenistic literary culture. The Pinakes cataloged 120 complete book rolls (a series of scrolls that form a complete book), which came to 200,000 scrolls in total.

Callimachus implemented Aristotle's classification of knowledge when creating his Pinakes. Table 12.1 provides readers with a detailed ordering of the areas of the Aristotelian classification system used by Callimachus. The Aristotelian classifications implemented in the Pinakes are natural philosophy (also known as theoretical sciences or natural sciences), practical sciences, and the productive sciences. Natural philosophy contains physics, astronomy, cosmology, meteorology, biology, botany, and the study of the soul, what we would call psychology today. Callimachus combined the

Table 12.1. Aristotelian classification.

Natural philosophy	*Less scientific*
Physics	Poetics
Astronomy	Rhetoric
Cosmology	Ethics
Meteorology	Politics
Studies of animals	Topics, categories
Study of plants	Interpretation
Study of soul	Constitutions
	(Law) lost
Analytics	
Metaphysics	

Source: G. W. Trompf, "The Classification of the Sciences and the Quest for Interdisciplinary: A Brief History of Ideas from Ancient Philosophy to Contemporary Environmental Science." *Environmental Conservation* 38, no. 2 (2011): 116, table 1.

Table 12.2. Aristotelian classification.

Poetry	*Prose*
Non-dramatic	History
Epic	Oratory
Elegiac	Philosophy
Iambic	Medicine
Melic	Law
Dramatic	Miscellany
Tragedy	
Comedy	

Source: Leo E. LaMontagne, *American Library Classification: With Special Reference to the Library of Congress*, (1961): 67.

practical and productive sciences into a category called the less scientific, which included politics, ethics, rhetoric, and the arts.

Like the cataloging of the Akkadians, Callimachus implemented similar practices in his catalog, by his use of the colophon and incipit cataloging systems. Additionally, within the Aristotelian classification, Callimachus "divided authors into classes and subclasses, arranged the authors alphabetically, added biographical data for each author, listed the titles written under each author, cited the opening words of each work, and listed the number of lines for each work."[24] He further divided these materials into two parts—poets and prosaists (writers of prose). Then he divided each of these into subcategories. Table 12.2 illustrates how Callimachus began to order materials, and subsequently, ideas, into a hierarchical order of importance.

FROM THE MIDDLE AGES TO THE ENLIGHTENMENT

Monastic Libraries

In contrast to the ancient Mesopotamian libraries that preserved every aspect of daily life, medieval libraries, specifically those found in monasteries, held small collections of books that covered very few subjects. The largest known monastic library collection, in Cluny, France, held fewer than 600 volumes.[25] Monastic libraries attempted to preserve literature of the ancient world as well as religious and contemporary works. Theodore W. Koch writes that:

> The medieval library catalogs list only a few hundred items at most . . . namely, the Christian writings and salvaged remains of antiquity—together with the learned and inspirational works . . . They show uniformly a triple division of their books into biblical writings, writings of the church fathers, and secular literature.[26]

The cataloging practices of the Alexandrian library did not carry on to the Middle Ages. Few monastic library catalogs were designed to be search tools, but were inventories, and they did not have meticulous classification schemes. They organized materials "based upon author, subject, and form."[27] The arrangements of these catalogs were hierarchical in nature and were three-tiered. They reflected the worldview of the monasteries they served. Scripture, the divinely revealed truth, sits on top of the hierarchy of medieval knowledge. The writings of the Christian Church fathers follow next, since they interpreted Scripture and provided moral teachings. Works of history, law (civil and canon law), and liturgy were often found in this second tier. Finally, the liberal arts serve as the third tier of the catalog. This tier contained works on grammar, rhetoric, poetry, dialectic, arithmetic, geometry, astronomy, and music.[28] This shaped the world in the medieval mind by reinforcing the belief that all knowledge, and therefore society, answers to Scripture and to ecclesiastical rules.

LATER CLASSIFICATION SYSTEMS

Outside of monastic libraries, scientists and philosophers categorized knowledge and the world around them into hierarchies. Using imagery from the Biblical creation story, many of these thinkers created symbolic trees of life or "omniscience," often called *arbor scientiae*, to depict the hierarchy of knowledge.[29] Medieval scholars used these *arbor scientiae* to help illustrate the scaffolding of knowledge of the natural world where the "structure of the universe . . . placed God above [the top most tree limbs], then celestial and angelic beings [within the branches], with humanity in the central stratum . . . and animals, insects, plants, rocks, and cavernous monsters and demons below."[30] These *arbor scientiae* would later be used to help Enlightenment scholars map out biological and evolutionary history. For instance, Athanasius Kircher (see figure 12.1) and Ernst Haeckel (see figure 12.2) benefitted by their use of the *arbor scientiae* to illustrate various categories of philosophical knowledge, scientific knowledge, and evolutionary biology. Their illustrations helped order the world according to their understanding of it, and depicted a homocentric worldview that placed humanity, specifically Europeans, at the top of the earthly creations.

LIBRARY OF CONGRESS AND THE LIBRARY OF CONGRESS CLASSIFICATION SYSTEM

In 1800, the Library of Congress was established by an act of Congress, to become a legislative library for the newly formed United States of America.

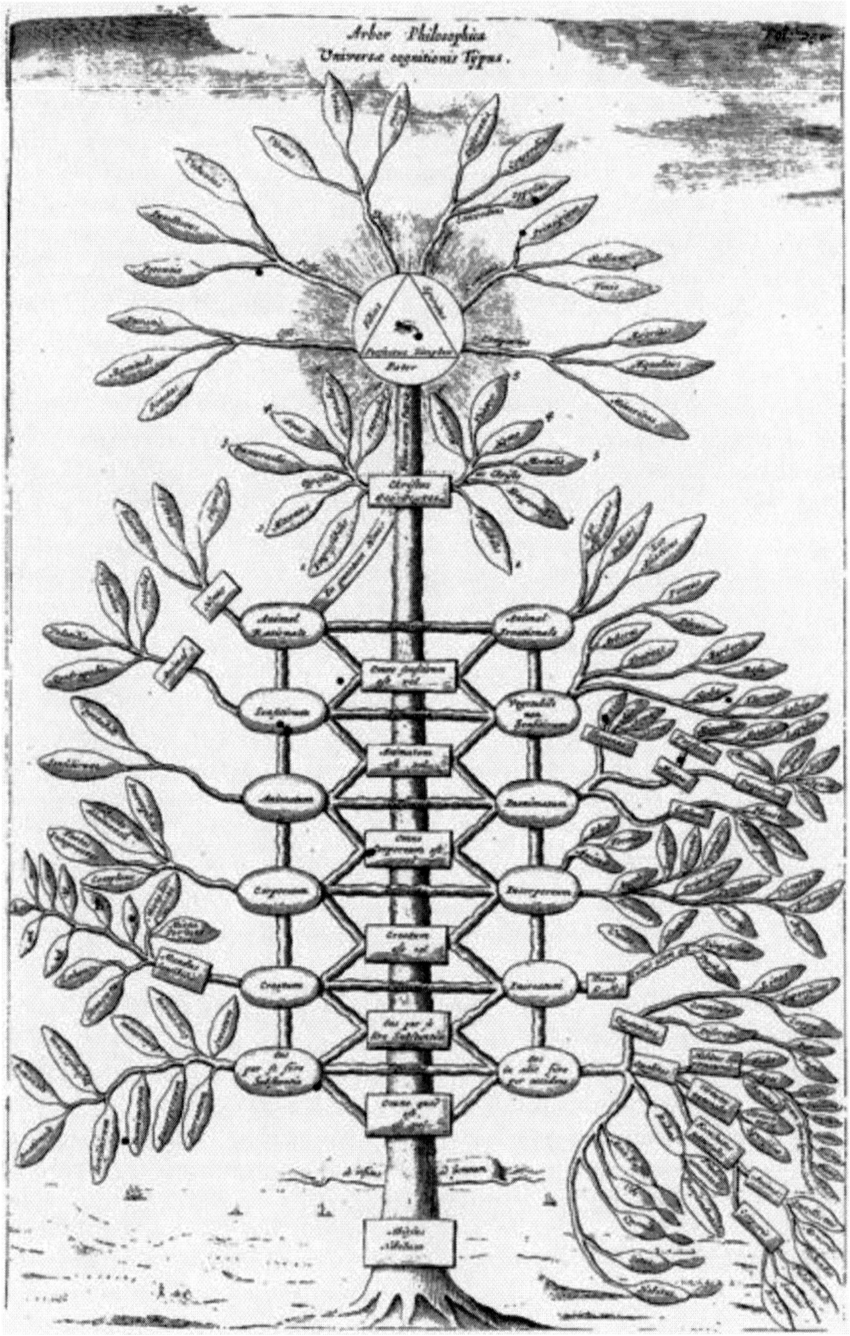

Figure 12.1. Kircher, Athanasius. "Philosophical Tree." Athanasius Kircher at Stanford, 2 Mar. 2011, Stanford University. Stanford, California, web.stanford.edu/group/kircher/cgi-bin/site/?attachment_id=795.

Image courtesy of the Department of Special Collections, Stanford University Libraries.

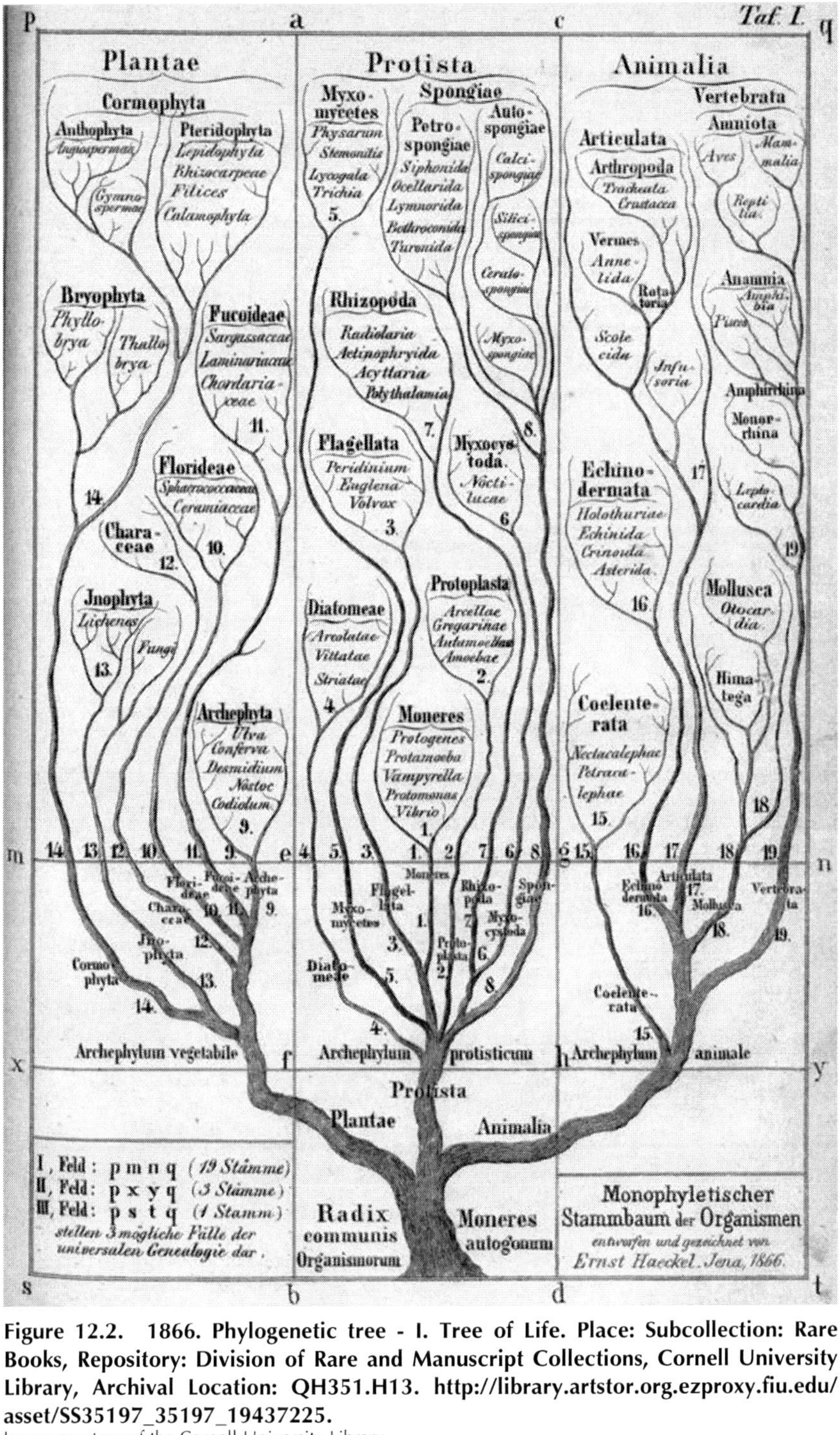

Figure 12.2. 1866. Phylogenetic tree - I. Tree of Life. Place: Subcollection: Rare Books, Repository: Division of Rare and Manuscript Collections, Cornell University Library, Archival Location: QH351.H13. http://library.artstor.org.ezproxy.fiu.edu/asset/SS35197_35197_19437225.

Image courtesy of the Cornell University Library.

After its destruction by the British army during the War of 1812, Congress laid the foundation for a new library with over 6,000 volumes of books donated from President Jefferson's private collection. Decades later, due to the increasing size of the library, it was rebuilt for the second time on a grander scale, and it reopened its doors in 1897. The Library of Congress serves as:

> A legislative library and the major research arm of the U.S. Congress; the copyright agency of the United States; a center for scholarship that collects research materials in many media and in most subjects from throughout the world in more than 450 languages . . . a government library that is heavily used by the executive branch and the judiciary . . . one of the world's largest providers of bibliographic data and products . . . a research center for the preservation and conservation of library materials; and the world's largest repository of maps, atlases, printed and recorded music, motion pictures, and television programs.[31]

In 1897, Librarian of Congress, John Russell Young, appointed catalogers and classifiers to emend the major shortcomings the library faced with its classification and cataloging systems. They gathered a great many subject specialists and classifiers to compile lists that would become the various classes of the new system. By 1901, Herbert Putnam, the next Librarian of Congress, made the final decision to adopt the current Library of Congress Classification System, which was more systematic and more functional.[32] This project produced 30 volumes that made up the Library of Congress Classification System, which was tailor-made for the library's holdings and reflected the collections.

Authors Zins and Santos conceived a map of human knowledge called the 10 Pillars of Knowledge. They write that human knowledge comprises 10 pillars, and these pillars serve as the basis of modern classification systems, such as the Library of Congress.[33] Table 12.3 provides detailed definitions for the 10 pillars.

As stated, the Library of Congress encompasses the 10 Pillars, organized into 21 basic classes, each identified by a single letter of the alphabet. Class A is assigned to general works. Next, classes B through P follow. They cover philosophy, psychology, religion, history, anthropology, the social sciences, the arts, literature, and law. Classes Q through V include the sciences and technology; and the classification system ends with bibliographies and library science in Class Z. Furthermore, the classes are divided into specific subclasses, identified by two-letter, or occasionally three-letter, combinations.[34] These subclasses include a loosely hierarchical arrangement of the topics pertinent to the subclass, going from the general to the more specific. Individual topics are often divided by geographic location, historical period, and bibliographic form.

Table 12.3. 10 Pillars of Knowledge.

Pillar	*Definition*
Foundations	The study of human knowledge.
Supernatural	Concerns mysticism and religion.
Matter and energy	Explores the basics of the physical world.
Space and earth	Explores our planet and outer space.
Nonhuman organisms	Explores the nonhuman living world, or fauna and flora.
Body and mind	Focuses on the human body and mind.
Society	Deals with the various aspects of human social life.
Thought and art	Studies the products of the human intellect and the arts.
Technology	Explores the products of human creativity, which are designed to achieve practical aims.
History	Encompasses human history.

Source: Chaim Zins and Placida Santos, "Mapping the Knowledge Covered by Library Classification Systems." *Journal of the American Society for Information Science and Technology* 62, no. 5 (2011). 88.

Classification Issues: Religion, Criminal Justice, and Race

Similar to the monastic and enlightenment classification systems, the Library of Congress reflected a worldview where Christianity, European philosophy, European history, and European law served as the foundation of the classification system, from which all other subjects branch away. Thus, from its inception, the Library of Congress Classification System has had its critics, who have argued that the subject headings reflect the biased and exclusionary views of those who dominate American society (i.e., those who self-identify as American, Western European, Protestant, white, heterosexual, and male) at various periods of its history.[35] Utilizing the language and perspective of the dominant classes, rather than implementing appropriate terminology, the Library of Congress unintentionally makes some library materials harder to find, while it stigmatizes minority groups with inaccurate, misleading, and often demeaning terms. These terms give the impression that the points of view of white, Anglo-American males are normal and preferable, while those of other groups, such as African-Americans, Native Americans, Hispanics, and LGBTQ people, are abnormal or are too insignificant to be considered.[36]

As an example of the criticisms described above, Library of Congress Class B covers philosophy, psychology, and religion. However, not every religion is equal under Class B. There is a bias towards Christian traditions. Christianity is assigned five subclasses, while other faith traditions are assigned only one, and often must share the same subclass with others, as if they are related faith traditions. Subclasses BR, BS, BT, BV, and BX encompass various aspects of the Christian faith and tradition. Subclass BR covers Christian literature, antiquities, martyrdom, dissent, liberalism, sacrilege, and biographies. Subclass BS covers the Bible, Biblical scholarship, and translations. BT and BV consist of Christian doctrinal theology and practical theology, respectively. BX covers most Christian denominations, their histories, and beliefs. Beyond the subclasses reserved for Christianity, Subclass BL covers mythology, cults, Hinduism, Jainism, and other religious traditions. The whole of Judaism is within Subclass BM, except for its Scriptures, which are combined with Christian Scriptures in BS. Islam and the Baha'i faith share Subclass BP with theosophy and other belief systems. For the patron who is unfamiliar with religious studies or with non-Christian traditions, this taxonomical framework may lead him or her to equate Hinduism with New Age cults. Similarly, the patron may come to think that Islam and the Baha'i faith are parts of the same religious tradition.

Other Library of Congress classifications have been blatantly racist and reflect the beliefs about race from earlier periods of American history. The system classified lynching under the subject heading of "Criminal Justice, Administration of," thus legitimizing the practice and providing a certain level of dignity to it. Berman writes:

> Is lynching somehow less "homicidal" or "murderous" than "strangling" or "poisoning"? Is it not an "offense against the person" usually resulting in a "violent death"? Are these qualities of lynching particularly diminished because the lynched, over the past century, have been largely black and the lynchers [sic] white?[37]

Moreover, Berman rightly states that the act of lynching was a terrorist technique used to maintain white supremacy and to subjugate African-Americans.

Classification Issues: LGBTQ Materials

The library, where information should be easily available, can also serve as a deterrent by unwittingly censoring LGBTQ materials and portraying LGBTQ persons as deviants. In the early 20th century, the Library of Congress used sexist, heterosexist, and homophobic terminology common for the period when categorizing materials pertaining to sexual minorities. Books related to homosexuality and other sexual minorities were classified under "Abnormal Sex Relations," thus enforcing society's anti-LGBTQ beliefs and forcing LGBTQ persons further into obscurity.

QUEERING THE LIBRARY OF CONGRESS

As discussed before, this chapter attempts to use queer theory to deconstruct the Library of Congress to analyze the who, why, and what that created its subject headings, and to challenge their assumptions. To successfully queer the Library of Congress, a shift is required from our traditional system—where only trained catalogers can create new categories and classification systems—to one that incorporates a more patron-generated or patron-sensitive system of classifying and organizing library materials.

Starting in the 1980s, there have been numerous changes to the Library of Congress Classification System, when librarians began to deconstruct the system and challenge many of its assumptions about society, race, and sexuality. Many offensive and biased subject headings were either deleted or changed. For instance, the subject heading "lynching" no longer falls under the category of criminal justice. It is now classified, along with homicide, murder, and terrorism, under HV, for criminology. Racial and ethnic classification terms have been updated as well. African-Americans are no longer classified as Negros. East and Southeast Asians and their cultures are no longer classified as Orientals or the Yellow Peril.[38]

There has also been a positive change in how LGBTQ materials are classified. Terms related to transgenderism, such as trans and transsexual, have been included in the classification system. The use of colloquial terms, such as gay and queer, now allows users to locate the materials they seek. With updated classifications based on colloquial terms, the Library of Congress is making LGBTQ materials more accessible to patrons without stigmatizing them. By using colloquial terms, the Library of Congress is in effect removing the censoring of library collections and removing obstacles for vulnerable patrons.

In order to continue in this positive direction, the Library of Congress should continue to queer itself by ending the practice of restricting who has authority to create new categories, especially in relation to topics that address minority groups. Therefore, the Library of Congress must consider consulting scholars and members of the specific minority groups to suggest appropriate terms and names for themselves. Library science scholars refer to this form of deconstruction as folksonomy.

Folksonomy is defined as:

> A method of categorizing information according to desire, taste, personal interest, communal knowledge, imagination, and so on. . . . [U]sers rename and redefine the placement of ideas and places according to an ever-shifting degree of categorization. Folksonomies provide reference systems, as any taxonomy does, but in digital spaces; they do so by challenging various assumptions about classifications and how such classifications generate meaning.[39]

This practice is common in social media applications and websites like Facebook and Twitter. Users may tag photos, locations, and events with a hashtag.

Similarly, on the website, LibraryThing, users digitally label books using tags based on their suggestions. Using this system, LibraryThing users can classify books pertaining to various subjects using terms that are familiar to the average reader. For instance, books pertaining to the transgender community will be tagged as transgender, trans, MTF, FTM, AFAB[40], or AMAB.[41] The members of individual minority groups will generate terms and classifications that are appropriate for them. These terms will no longer hold the sting of heterosexism or racism but will instead reflect the views and values of the group. Who better to classify information about the transgender community than individuals in that community? As was already discussed, when library catalogers attempted to create subject terms for books pertaining to minority communities, they often used terms that were inappropriate and insensitive, based on personal beliefs, bias, and the prejudice of the time. Allowing folksonomy into the classification process will help prevent future offensive terms and classifications to be created for minority groups.

The current library classification of African-Americans is an example of folksonomy. For much of the 20th century, African-Americans were classified as Negros. In the 1970s, the term was changed to Blacks, and in the 1990s, it was briefly changed to Afro-Americans before it was amended to African-Americans. This change was instituted in part by the inclusion of African-Americans in the work of classification, which in turn was a challenge to the old system of classification and a deconstruction of the who, why, and what that produced older classifications and subject headings.

CONCLUSION

Libraries and librarians have been classifying human knowledge for millennia, starting with agricultural records and clay tablets, and moving to categorization of various aspects of human culture, including belief systems, sexuality, and race. With the advent of the Library of Congress Classification System, the classification and organization of information unwillingly contributed to how patrons and society as a whole perceive the world. Racist and heterosexist terminology used in the system helped preserve and enforce unjust systems. Since the 1980s, librarians have been challenging troublesome classification terms and exploring new ways to improve how knowledge is classified and organized in libraries. By exploring the history of libraries and classification systems, one will better appreciate how queering the Library of Congress has led to the improvement of these classification systems, by consulting representatives of various groups to create and institute proper terminology pertaining to the various segments of human societies.

WORKS CITED

Bales, Stephen E. "Tracing the Archetypal Academic Librarian." *Library Philosophy and Practice* (2009). *Academic OneFile.*

Berman, Sanford. *Prejudices and Antipathies: A Tract on the LC Subject Heads Concerning People*. Metuchen, NJ: The Scarecrow, Press, Inc., 1971.

Boda, Istvan, Miklos Benyei, and Erzsebet Toth. "New Dimensions of an Ancient Library: The Library of Alexandria." Presentation at the 4th IEEE International Conference on Cognitive Infocommunications, Budapest, Hungary, December 2–5, 2013.

Casson, Lionel. *Libraries in the Ancient World*. New Haven: Yale University Press, 2001.

Cole, John Young. *Jefferson's Legacy: A Brief History of the Library of Congress*. Washington: Library of Congress, 1993.

Dilley, Patrick. "Queer Theory: Under Construction," *International Journal of Qualitative Studies in Education* 12, no. 5 (1999): 457. Taylor & Francis Online.

Haeckel, Ernst. *1866. Pyhlogenetic tree*—I. Tree of Life. Place: Subcollection: Rare Books, Repository: Division of Rare and Manuscript Collections, Cornell University Library, Archival Location: QH351.H13. http://library.artstor.org.ezproxy.fiu.edu/asset/SS35197_35197_19437225

Jenkins, Christine. "Gay and Lesbian Issues for School Libraries and Librarians." *Gay and Lesbian Library Service* (1990): 11–24.

Kircher, Athanasius. "Philosophical Tree." *Athanasius Kircher at Stanford*, 2 Mar. 2011, Stanford University. Stanford, California, web.stanford.edu/group/kircher/cgi-bin/site/?attachment_id=795.

Knowlton, Steven A. "Three Decades Since Prejudices and Antipathies: A Study of Changes in the Library of Congress Subject Headings." *Cataloging & Classification Quarterly* 9, no. 3 (July 1939): 123–45. Tandfonline (JSTOR).

Koch, Theodore W. "New Light on Old Libraries." *Library Quarterly* 4, no. 2 (April 1934): 244–52. JSTOR.

LaMontagne, Leo E. *American Library Classification: With Special Reference to the Library of Congress*. Hamden, Conn: Shoe String Press, 1961.

Lerner, Frederick Andrew. *The Story of Libraries: From the Invention of Writing to the Computer Age*. New York: Continuum, 1998.

"Library of Congress Classification." Library of Congress Classification. Accessed July 24, 2015. http://www.loc.gov/catdir/cpso/lcc.html.

Marshall, Joan. "LC Labeling: An Indictment" In *Revolting Librarians*. San Francisco: Booklegger Press, 1972.

Rice, Jeff. "Folksono(me)." *JAC* 28, no. 1/2 (2008): 181–208. JSTOR.

Rubin, Richard. *Foundations of Library and Information Science*. New York: Neal-Schuman Publishers, 2010. Print.

Scrivner, Buford. "Carolingian Monastic Library Catalogs and Medieval Classification of Knowledge." *The Journal of Library History (1974–1987)* 15, no. 4 (1980): 427–44. JSTOR.

Trompf, G. W. "The Classification of the Sciences and the Quest for Interdisciplinary: A Brief History of Ideas from Ancient Philosophy to Contemporary Environmental Science." *Environmental Conservation* 38, no. 2 (2011): 113–26.

Zins, Chaim, and Placida Santos. "Mapping the Knowledge Covered by Library Classification Systems." *Journal of the American Society for Information Science and Technology* 62, no. 5 (2011): 877–901.

NOTES

1. Patrick Dilley, "Queer Theory: Under Construction," *International Journal of Qualitative Studies in Education* 12, no. 5 (1999): 457.
2. Ibid., 457–458.
3. Ibid., 458.
4. Ibid., 459.
5. Ibid., 460.
6. Frederick Andrew Lerner, *The Story of Libraries: From the Invention of Writing to the Computer Age* (New York: Continuum, 1998), 13.
7. Ibid., 14.
8. Ibid.
9. Ibid.
10. Ibid., 15.
11. "colophon, n." OED Online. December 2016. Oxford University Press. http://www.oed.com/view/Entry/36552?rskey=CrALWu&result=2 (accessed January 28, 2017).
12. Lionel Casson, *Libraries in the Ancient World* (New Haven: Yale University Press, 2001), 5.
13. Ibid.
14. Lerner, *The Story of Libraries*, 15.
15. Ibid.
16. Ibid.
17. Casson, *Libraries in the Ancient World*, 9.
18. Lerner, *The Story of Libraries*, 15.
19. Casson, *Libraries in the Ancient World*, 5.
20. Richard Rubin, *Foundations of Library and Information Science* (New York: Neal-Schuman Publishers, 2010), 36.
21. Theodore W. Koch, "New Light on Old Libraries," *Library Quarterly* 4, no. 2 (1934): 245.
22. Stephen E. Bales, "Tracing the Archetypal Academic Librarian," *Library Philosophy and Practice* (2009): 4.
23. Istvan Boda, Miklos Benyei, and Erzsebet Toth, "New Dimensions of an Ancient Library: The Library of Alexandria" (presentation, IEEE International Conference on Cognitive Infocommunications, Budapest, Hungary, December 2–5, 2013), 537.
24. Bales, "Tracing the Archetypal Academic Librarian," 6.

25. Buford Scrivner, "Carolingian Monastic Library Catalogs and Medieval Classification of Knowledge." *The Journal of Library History (1974–1987)* 15, no. 4 (1980): 428.

26. Koch, "New Light on Old Libraries," 246–47.

27. Scrivner. "Carolingian Monastic Library Catalogs and Medieval Classification of Knowledge," 433.

28. Ibid., 437.

29. Trompf, "The Classification of the Sciences and the Quest for Interdisciplinary," 117–18.

30. Ibid., 118.

31. John Young Cole, *Jefferson's Legacy: A Brief History of the Library of Congress* (Washington: Library of Congress, 1993), 11–12.

32. Leo E. LaMontagne, *American Library Classification: With Special Reference to the Library of Congress* (Hamden, Conn: Shoe String Press,1961), 234.

33. Chaim Zins and Placida Santos, "Mapping the Knowledge Covered by Library Classification Systems." *Journal of the American Society for Information Science and Technology* 62, no. 5 (2011), 88.

34. "Library of Congress Classification." Library of Congress Classification. Accessed July 24, 2015. http://www.loc.gov/catdir/cpso/lcc.html.

35. Joan Marshall, "LC Labeling: An Indictment" in *Revolting Librarians*. (San Francisco: Booklegger Press, 1972), 46.

36. Steven A. Knowlton, "Three Decades Since Prejudice and Antipathies: A Study of Changes in the Library of Congress Subject Headings." *Cataloging & Classification Quarterly* 9, no. 3 (July 1939): 125.

37. Sanford Berman, *Prejudices and Antipathies: A Tract on the LC Subject Heads Concerning People* (Metuchen, NJ: The Scarecrow, Press, 1971), 67.

38. Knowlton, "Three Decades Since Prejudice and Antipathies," 130.

39. Jeff Rice. "Folksono(me)." *JAC* 28, no. 1/2 (2008): 181–208. JSTOR.

40. Assigned female/male at birth.

41. Assigned male/female at birth.

Petrichor: An Afterword

Timothy Morton

In this book's introduction, we find the following sentence: "[Eve Sedgwick] notes that of the many ways we might differentiate genital activity—preference for certain acts, sensations, frequency relations of power and so forth 'precisely one, the gender of the object choice' remains '*the* dimension' by which we understand 'sexual orientation.'"[1] Isn't it the case that the list of phenomena here are not to do with identity—with the essence of an entity such as "woman" or "man"—but with *data.* A sensation is a kind of data. A relation of power is experienced as data. An act is how a being—a system, an individual, a group . . . —is behaving in the moment, and so an act is also a kind of data.

Data means what is given (Latin, *dare*, to give). Modern science is founded on a strict division between things and data about those things. David Hume restricted knowledge to data, and Kant explains why: there is a transcendental—which means irreducible, so that you can't point to it in empirically measurable extensional spacetime—gap between a thing and data "about" that thing. Kant uses the example of raindrops. Raindrops are wet, spherical, they have a certain size, a certain momentum, and so on. These are all examples of raindrop data. At no point do our encounters with the raindrop engage the raindrop as such. Yet a raindrop manifests exactly these data. Raindrops are not gumdrops (more's the pity). In his example—Kant was afraid of examples, because he intuited how they might wriggle out of control—even the mathematical properties of the raindrop, its extensionality in spacetime, are not pointing to the actual raindrop. This is despite how Kant himself tried to stabilize the shockwave of his grounding of Hume in the thinking (human) mind, and in particular that mind's ability to mathematize: to measure various kinds of extension.

In this weirdly boring sense, Kant was a sizeist. The length and movement of a thing is all that matters. But in the example, even the physical dimensions and speeds of a thing aren't the "reality" of that thing as opposed say to their smell. Petrichor, the smell of dust after rain, is just as much raindrop data as various equations involving pi that might plot their spherical properties. Reality isn't sizeist.

In order to have data of this kind, there must be some kind(s) of data-ness, in other words, all kinds of sensuality must be intrinsic to reality, and the name one might give these sensualities is givenness. Petrichor will be my example of this givenness here, a givenness not unrelated etymologically to the dust or earth (*adamah*) from which God fashions the human, as Zairong Xiang's essay explores. I am fond of brutalist architecture, but not for anything elaborate that it appears to be saying to the buildings around it or to the architectural lineages that structure that field. I like brutalism because its buildings have that wet concrete smell: they are massive homages to petrichor.

I also use petrichor to destabilize the lookism inherent in science, religion, and philosophy. *I see* is the metaphor of choice for knowing, not *I smell*. Sight has been implicated for several thousand years in the default ontology that places (human) subjects "over here" and everything else "over there," which is why it takes Maurice Merleau-Ponty pages and pages of rhetorical gyration to get at the physical qualities of seeing, the fact that seeing happens in the wet red flesh of your eye, that your field of vision is "palpating" the "object."[2] Which implies, by the way, that the "object" is always already palpating you. Smell is far more obviously a manifold of entities entwined together: noses, concrete, rain, brains, sinuses . . . I have a particularly acute sense of smell, and if I hadn't become a scholar, I might have become a "nose" for the perfume industry. So this afterword will cleave to that mode of givenness as far as it can.

The well-worn logic of using sight as a way to evoke knowing or gnosis is precisely to objectify, to de-sensualize, to make a sharp difference between something and nothing that is the basis of a bleak nihilism. What would happen if we smelled truth? What if we took very seriously the non-visibility inherent in the esoteric Buddhist tantras that talk of searching for the nature of mind, only to relax in a *nothing to see*? Nothing to see, but in a way, perhaps something to smell? The way in which the notion of *observation* is troubled in quantum theory, the most consistent way physics knows to assess reality at present, speaks precisely to this *nothing to see*, as Fern Feldman's essay makes clear.

Some theologians argue that givenness is what the sacred consists of. The way in which, in the Rothko Chapel down my street, the petrichor and smell of the paint on the canvases—Rothko paintings have a strangely fresh smell, as if they were just painted, especially in the humidity of Houston, Texas—envelops visitors. I do not encounter the interior of the Rothko chapel as a gallery in which objectified artworks are suspended in a deliberately antiseptic "white cube" so as to distance them and reduce the human viewer to their eye. When I enter the Rothko Chapel, it's as if I find myself permeated with a givenness I have sometimes been keeping at bay outside the building's confines.

Givenness always manifests as contingent and in motion, constantly evaporating and shifting like a specter, like petrichor.

In adjusting to their discipline, scientists often make big experiential stretches. Geology Ph.D.s, for example, report that at first the prospect of encountering familiar objects such as stones at non-anthropocentric scales can be really daunting. One becomes aware of billions of years' worth of history without a human in sight. When tested by psychology, most people are able to narrate the correct sequence of events: Big Bang, stars, dust and gas cloud, planets, Earth, Moon, oceans, evolution . . . But the duration of those events remains uncertain for most people. It's in the field of duration that the big stretch is required of geologists. The smell of dust after rain is the smell of billions of years.

Science sniffs out patterns in data, and therefore givenness is its logical possibility condition. This means that there is a resonant affinity between science and religion, if givenness is the realm of the sacred. The playground of science is this sacred realm. In a sense, science has more in common with art appreciation than with scientism, which is a mode of belief in scientific factoids and in assumed ontological implications of science, which is said, for instance, to include a narrowly defined form of atheism.

I will now begin to swap the term *religion* for the more slippery, riskier term *spirituality*. I am trying to signal, with this swap, something that inheres in the study and practice and rhetoric of religion that must be akin to the difference within the practice and study and rhetoric of science, the difference between science and *scientism*. Scientism and religion (the etymology of the term in the notion of binding back) seek to stabilize science and spirituality. I will show too briefly why this is impossible: deeply impossible, in the same way that having a square circle would be deeply impossible. A kind of contradiction that cannot be accepted operates here. Strangely, though perhaps not so strangely, a logic that accepts that some contradictions can be true structures the fields of science and spirituality, in ways that I claim are related precisely through the notion of queerness. And this is why religion and scientism, or as Lisa L. Stenmark puts it in this volume, "a belief in a single truth, a single understanding of it, and the existence of a select group with access to that truth and that understanding," what Whitney Bauman calls "the perils of authority," distort what is common to science and spirituality.

By refusing to make an ontology out of patterns in data, science never seeks to stabilize reality. In particular science never stabilizes reality around a notion of identity. The phenomenology of science—how it arises—is like the mythical garuda, who when they hatch, takes off immediately into outer space. The phenomenology of science, to put it another way, is outrageous.

Moreover, the inevitable queerness of science as opposed to scientism—and the queerness of spirituality as opposed to religion—is enabled by the queerness of the real. Perhaps the key to grasping this is accepting the notion that movement is hardwired into the real, not some superficial or accidental or illusory quality of it. Consider the quantum of *movement*—a vibrant hesitation, a quivering one can now glimpse in objects cooled close to absolute zero and isolated in a vacuum. Nothing is pushing them, no mechanism is moving them from A to B, by definition. Yet objects isolated in this way are observed to be *quivering*: to be still and moving at the very same time, a "performativity of deviance" (Julia Watts Belser) default to being an entity at all.[3]

It's skillful of Philip Clayton and Kirianna Florez to have titled their essay on the movement called gender fluidity "Both/And," because this gets at the fundamental paradox of movement; at bottom, movement is not a mechanical traversal of a thing "in" extensional space from A to B. Fluidity, they argue, is structural to gender as such. Gender cannot be stabilized, because givenness—in their terms, phenotypical expressions of the genome—cannot be stabilized, because the real just isn't stable: lack of stability is its intrinsic, defining characteristic. It is by definition what Carol Wayne White calls a *bastard* phenomenon. Science—an attunement to what is given, to data—and spirituality—an attunement to givenness, to what structures the field of the given—are queer in related ways, because the real of which data and givenness flow, is queer.

To put it in the terms of Teresa Hornsby's essay, there is a fundamental ambiguity in the worlds we inhabit because there is a fundamental ambiguity in the real. This is why ambiguity is often a signal of accuracy, an accuracy one can sense in the deepening ambiguities inherent in smelling, where increased sensibility results in increased and therefore more disparate data, making the boundary between (say) pleasant and unpleasant neither thin nor rigid. A scent profoundly troubles the notion of self-identity: that to be a thing is to be consistent all the way through, as if every biological aspect of a human being had a kind of "Intel Inside" sticker on it, inscribed with the words *This is a human being component*. A whole smell is always less than the sum of its parts. The bacterial microbiome contributes to how a human smells.

What is required to think the beings that science and spirituality detect is a weird holism that does not gather things together as religion too often does, as parts totally subsumed in a despotic whole that makes their specificity ultimately meaningless, because each part is now only a component in a machine-like entity that can replace them as necessary. In this volume, Alex Carr Johnson gets at the destructive qualities of such thoughts in ecological discourse. Progressive thought requires holism, not reductionism. If queer thought can contribute anything to holism, it would be drastically to reimagine it as something like what happens in the petrichor phenomenon: a smell that evokes both dust and rain, dry and wet, two parts of a whole whose drastic difference signals a way for wholes to be less than the sum of their parts, a lessness that provides wiggle room for things to happen, and therefore for change to occur (physical, political, psychological . . .). A whole is always some kind of *emulsion* in which parts are not totally dissolved, a library bursting with contradictory (and dusty) books that cannot be contained in a holism that dissolves them with a singular and violent archival procedure, as Carlos R. Fernández explores here.[4]

In conclusion, we are left with the fact that the very ideas of science and religion are kinds of queerness. Queerness in this sense cannot be thought simply as conceptual or, for that matter, physical openings or transcendings, but rather as forms of openness intrinsically endowed with qualities. Sensing this vividly in her essay here, Emilie M. Townes endows the very term queer with the sensual qualities of the American South by using the word *quare*. Queerness isn't bland (in the language of racism, white) vanilla openness. There can be no such thing. It's rather like how, in Buddhism, emptiness doesn't mean that Buddha can't use a toilet or fix herself a plate of sliced fruit or take a walk through some brutalist architecture, the damp concrete charming her nose. What is evoked in the Genesis term *tĕhôm* as it resonates in Hornsby's essay is precisely a sensual void, not simply a bland space. An *incarnate* void, to put it in the terms of Catherine Keller and Laurel Schneider. A void perfumed with wet dust.

An evocative openness to what is evocatively open structures both human fields, because they are related, and because the real from which they draw their data and their givenness are evocatively open. The spaciousness is sensual, like the smell of dust after rain: petrichor.

WORKS CITED

Merleau-Ponty, Maurice. 1968. *The Visible and the Invisible (Studies in Philosophy and Existential Phenomenology)*. Translated by Alphonso Lingis and edited by Claude Lefort. Evanston, IL: Northwestern University Press.

Morton, Timothy. 2017. *Humankind: Solidarity with Nonhuman People*. London, UK: Verso.

Safavi-Naeini, Amir H., Jasper Chan, Jeff T. Hill, T. P. Mayer Alegre, Alex Krause, and Oskar Painter. 2012. "Observation of Quantum Motion of a Nonomechanical Resonator." *Physical Review Letters,* 8 no. 3: 033602. (Epub).

Sedgwick, Eve K. 1990. *Epistemology of the Closet.* Berkeley, CA: University of California Press.

NOTES

1. Eve K. Sedgwick, *Epistemology of the Closet* (Berkeley: University of California Press, 1990), 8.

2. Maurice Merleau-Ponty, *The Visible and the Invisible (Studies in Philosophy and Existential Phenomenology)*, tr. Alphonso Lingis, ed. Claude Lefort (Evanston: Northwestern University Press, 1968).

3. Amir H. Safavi-Naeini, Jasper Chan, Jeff T. Hill, T. P. Mayer Alegre, Alex Krause, and Oskar Painter. 2012. "Observation of Quantum Motion of a Nanomechanical Resonator." *Physical Review Letters*, art. 033602, January 17, 2012.

4. Space is lacking for a full logical proof of this new form of holism, which elsewhere I call subscendence. See Timothy Morton, *Humankind: Solidarity with Nonhuman People* (London: Verso, 2017).

Annotated Bibliography

Aside from important primary texts—e.g., Jacques Derrida, Michel Foucault, and Judith Butler as well as from individual bibliographies in this volume—the following provide excellent overviews of queer theory, the religion and science discourse, and related critical discourses.

QUEER THEORY

Cox, Matthew B. and Michael J. Faris. "Annotated Bibliography of LGBTQ Rhetorics." (peer-reviews online journal). *Present Tense: A Journal of Rhetoric in Society* 4 no. 2 (2015). http://www.presenttensejournal.org/volume-4/an-annotated-bibliography-of-lgbtq-rhetorics/ Somewhat specialized, but a thorough bibliography, with a good discussion of LGBT issues and Queer Theory.

Eng, David L., Judith Halberstam, and José Esteban Muñoz, eds. "What's Queer about Queer Studies Now?" Special Issue of *Social Text* 84–85 23/3–4 (Fall-Winter 2005). Various authors' reflections on the need to keep queering queer theory in light of its institutionalization.

Jagose, Annamarie. *Queer Theory: An Introduction.* New York, NY: New York University Press, 1997. Accessible, provides excellent context with discussion of gay/lesbian movement and background to queer studies, along with good discussions of Derrida and Foucault. Discussion of actual queer theory is somewhat limited (esp. race, class, gender, etc. issues), but covers Butler, Halperin et al.

Marcus, Sharon. "Queer Theory for Everyone: A Review Essay." *Signs: Journal of Women in Culture and Society* 31, no. 1 (Autumn 2005): 191–218.

Reed, T.V. *CulturalPolitics.net.* Provides excellent resources—including an annotated bibliography and links to further resources. Covers queer theory, as well as postmodern, poststructural, and postcolonial theories.

Warner, Michael, ed. *Fear of a Queer Planet: Queer Politics and Social Theory*. Minneapolis: University of Minnesota Press, 1993. A pioneering book that pushes queer theory to become something other than "gay and lesbian studies" and into social, economic, and transcultural theory. Introduction is key reading.

Wilchins, Riki. *Queer Theory, Gender Theory: An Instant Primer*. Los Angeles: Alyson Books, 2004. Straightforward, readable introduction to queer theory, including discussion of Foucault, Derrida, and Butler. Includes anecdotes from her own activism, along with concrete examples from LGBT and feminist politics.

QUEER SCIENCE

Begemihl, Bruce. *Biological Exuberance: Animal Homosexuality and Natural Diversity*. New York: St. Martin's Press, 1999. Comprehensive discussion of over 250 years of biological science that reveals an "exhuberant variety" of sexual expression among animals—mammals, birds, reptiles, insects—including same-sex courtship, pair-bonding, non-reproductive sex, co-parenting and lifelong homosexual.

Fausto-Sterling, Anne. *Sex-Gender: Biology in a Social World*. New York: Routledge, 2012. Accessible discussion of the biological and cultural underpinnings of gender, extends discussion into moral, ethical, social, and political considerations.

Roughgarden, Joan: *Evolution's Rainbow: Diversity, Gender and Sexuality in Nature and People*. Berkeley, CA: University of California Press, 2004. This is a very accessible introduction to "queer" identities throughout the natural world. Furthermore, Roughgarden argues that heteronormative interpretations of evolution have placed too much emphasis on competition and not enough emphasis on cooperation.

Sandilands, Catriona and Bruce Erickson. *Queer Ecologies: Sex, Nature, Politics Desire*. Bloomington, IN: Indiana University Press, 2010. This edited volume addresses the intersection of queer theory with environmental sciences and environmental studies.

QUEER RELIGION

Armour, Ellen T. "Blinding Me with (Queer) Science: Religion, Sexuality, and (Post?) Modernity." *International Journal for Philosophy of Religion* 68, no. 1/3. Special Volume on the Occasion of the 40th Anniversary of This Journal (December 2010): 107–119. A very good discussion of Foucault, Butler, and the science of (homo)sexuality. Armour suggests that "premodern" perspectives (e.g., within the history of Christianity) can challenge controversies over homosexuality within global Christianity.

Boisvert, Donald and Jay E. Johnson. *Queer Religion: Homosexuality in Modern Religious History, 2 vols*. Santa Barbara, CA: ABC-CLIO, 2012. This edited 2-volume series contains chapters on queer identities and queer concepts that exist throughout the world's religious traditions.

Schippert, Claudia "Implications of Queer Theory for the Study of Religion and Gender: Entering the Third Decade." *Religion and Gender* 1, no. 1 (2011): 66–84. Explores the conceptual and contextual shifts in queer theory—especially transnational and diasporic queer scholarship, and shifts in conceptions of agency—and its implications for the study of religion and gender.

RELIGION AND SCIENCE

Barbour, Ian. *When Science Meets Religions: Enemies, Strangers or Partners*. New York: Harper Collins, 2000. Written by who many consider to be the "father" of contemporary, western discussions of "science and religion," this text is a great introduction to Barbour's four main models for relating the two: conflict, non-overlapping magesteria, dialogue, and integration.

Ferngren, Gary B. ed. *Science and Religion*. Baltimore: Johns Hopkins University Press, 2002. This collection of essays provides the reader with a good, historical introduction to the relationship between (mostly monotheistic) religions and science.

Matthews, Clifford, Mary Evelyn Tucker, and Philip Hefner, eds. *When World's Converge: What Science and Religion Tell Us about the Story of the Universe and Our Place in It.* Peru, IL: Carus Publishers, 2002. This collection of essays explores contemporary scientific cosmology, ecology, and modern science (in general) as they relate to many of the different, extant world religions. It is a very helpful text in a field that usually has focused on western monotheisms.

Numbers, Ronald. *Galileo Goes to Jail and Other Myths about Science and Religion.* Cambridge, MA: Harvard University Press. 2009. This book does an excellent job of exploring historical misinterpretations about the relationship between science and religion as only one of conflict. He takes the reader beyond the conflict model toward a more complex understanding of the relationship.

OTHER CRITICAL DISCOURSES

Anzaldúa, Gloria. *Borderlands/La Frontera: The New Mestiza*. San Francisco: Aunt Lute Books, 1987. Groundbreaking exploration of identity on the border between geographic, linguistic, cultural, and sexual cultures. Using a hybrid style—poetry, prose, and personal essays; English, Spanish, and (occasionally) Nahuatl—Anzaldúa describes a hybrid consciousness that changes perception of reality and tolerates contradictions and ambiguity.

Boisvert, Donald and Jay E. Johnson. *Queer Religion: Homosexuality in Modern Religious History, 2 vols*. Santa Barbara, CA: ABC-CLIO, 2012. This edited 2-volume series contains chapters on queer identities and queer concepts that exist throughout the world's religious traditions.

Collins, Patricia and Sirma Bilge. *Intersectionality (Key Concepts).* Cambridge, MA: Polity Press, 2016. Comprehensive but readable introduction to intersectionality, with historical background. Explores relevance of intersectional analysis for a number of topics.

Crenshaw, Kimberlé, ed. *Critical Race Theory: The Key Writings That Formed the Movement*. New York: New Press, 1995. Brings together key writings from the two disciplines that form the core of critical race theory: sociology (and the social construction of race) and critical legal studies.

Delgado, Richard and Jean Stefanic. *Critical Race Theory: An Introduction.* 2nd ed., revised edition. New York: NYU Press, 2012. The definitive discussion of Critical Race Theory; straightforward, brief, but comprehensive introductory text with good examples.

hooks, bell. *Feminist Theory: From Margin to Center.* 2nd ed. Boston: South End Press, 2000. A classic text that moves women of color feminism into the center of feminist thought.

Lorde, Audre. *Sister Outsider: Essays and Speeches.* Freedom, CA: Crossing Press, 1984. Collection of Lorde's work theorizing intersectional queer, womanist, class-based position. Enjoyable to read, grounded in practical, experiential politics.

Young, Robert. *Postcolonialism: An Historical Introduction.* New York/London: Blackwell, 2001. Comprehensive introduction to postcolonial theory, good historical stuff.

Index

About the Contributors

Whitney Bauman is Associate Professor of Religious Studies at Florida International University in Miami. He teaches and lectures on science and religion, religion and nature, and religion and queer theory. His books include *Religion and Ecology: Developing a Planetary Ethic* (Columbia University Press, 2014), *Theology Creation and Environmental Ethics* (Routledge, 2009), with Lucas Johnston, *Science and Religion: One Planet, Many Possibilities* (Routledge, 2014), and with Kevin O'Brien and Richard Bohannon, *Grounding Religion: A Fieldguide to the Study of Religion and Ecology* (Routledge, 2010). He is currently working on a manuscript that examines the religious influences on Ernst Haeckel's understanding of the natural world.

Julia Watts Belser is Associate Professor of Jewish Studies in the Theology Department at Georgetown University. She studies rabbinic Jewish culture in late antiquity and contemporary Jewish feminist ethics, with particular interests in disability studies, gender and sexuality, and environmental ethics. Belser is the author of *Power, Ethics, and Ecology in Jewish Late Antiquity: Rabbinic Responses to Drought and Disaster* (Cambridge University Press, 2015). Her next book, *Fashioning Catastrophe: Gender, Sexuality, and Disability in the Ruins of Jerusalem,* is forthcoming from Oxford University Press.

Philip Clayton is Ingraham Professor at Claremont School of Theology and an affiliated faculty member at Claremont Graduate University. He received a joint doctorate in philosophy and religious studies from Yale University and has published some 24 books, largely in theology and philosophy, comparative theologies, and the religion-and-science field. These include *Mind and Emergence; The Reemergence of Emergence; In Quest of Freedom:*

The Emergence of Spirit in the Natural World; and *The Oxford Handbook of Religion and Science*.

Rabbi Fern Feldman is a singer, scholar, spiritual counselor, and service leader. She was ordained in 2003 by Aleph: Alliance for Jewish Renewal. Her academic background also includes degrees in nursing and French Literature. Her interests include exploring the sacred dark in Jewish tradition, non-binary approaches to gender in Kabbalah, and reading quantum physics and Kabbalah through one another. Her article, "The Burning Bush and Black Fire" was recently published in *Tikkun.* She also works part time as a nurse practitioner and clinical hypnotherapist, and lives in Santa Cruz, CA. Her website is www.rabbifernfeldman.com.

Carlos R. Fernández is the International Documents Librarian at Florida International University's Steven J. and Dorothea Green Library. He serves as the library's liaison to multiple departments at the Green School for International and Public Affairs and is responsible for the university's United Nations and European Union depositories. Fernández holds master's degrees in Library and Information Science from Florida State University (2012), theology from Chicago Theological Seminary (2008), and a Bachelor of Arts in anthropology from the University of Miami (2003).

Kirianna Florez is a doctoral student in Hebrew Bible at Claremont School of Theology. Her research interests include Gender and Sexuality, Children and Childhood Theory, ANE Religions and Literature, Religion and Pop Culture, and Religion and Literature.

Teresa Hornsby is a Professor of Religion at Drury University. Dr. Hornsby holds an MTS degree from Harvard Divinity School, and an MA and Ph.D. in New Testament Studies from Vanderbilt University. Her most recent book, *Transgender, Intersex, and Biblical Interpretation,* co-authored with Deryn Guest, was published in July 2016. She is a co-editor (with Ken Stone) of *Bible Trouble: Queer Reading at the Boundaries of Biblical Scholarship,* and author of *Sex Texts from the Bible,* as well as numerous essays, chapters, and encyclopedia entries. Dr. Hornsby's hobbies are playing drums, brewing beer, and restoring vintage drum kits.

Alex Carr Johnson is a writer and conservationist who lives in the mountains of western Colorado. He currently serves as Executive Director of the Western Slope Conservation Center. His writing has appeared in *Orion* magazine, *Earth Island Journal*, *High Country News*, *Astrobiology*, and elsewhere. He earned his Masters of Science in Environmental Studies from the University of Montana and is a Wyss Scholar for Conservation in the Intermountain West. Alex is currently working on a memoir that follows his and his husband's search for home and family as told through their various journeys into wild landscapes.

Catherine Keller is George T. Cobb Professor of Constructive Theology in the Graduate Division of Religion of Drew University. Books she has authored include *From a Broken Web: Separation, Sexism and Self, Apocalypse Now & Then; God & Power; Face of the Deep: A Theology of Becoming; On the Mystery: Discerning God in Process; Cloud of the Impossible: Negative Theology and Planetary Entanglement;* and *Intercarnations: Exercises in Theological Possibility.* She has co-edited several volumes of the Drew Transdisciplinary Theological Colloquium, including *Postcolonial Theologies; Ecospirit; Apophatic Bodies*; *Polydoxy; Common Good/s: Ecology, Economy and Political Theology*; and *Entangled Worlds: Religion, Science and the New Materialism.*

Timothy Morton is the Rita Shea Guffey Chair in English at Rice University. His works include: *Humankind: Solidarity with NonHuman People* (Verso, 2017); *Dark Ecology: For a Logic of Future Coexistence* (Columbia University Press, 2016); *Nothing: Three Inquiries in Buddhism and Critical Theory* (Chicago, 2015); *Hyperobjects: Philosophy and Ecology after the End of the World* (Minnesota, 2013); and *Ecology without Nature* (Harvard, 2007), nine other books, and one hundred and twenty essays on philosophy, ecology, literature, food, and music. He blogs regularly at http://www.ecologywithoutnature.blogspot.com.

Laurel Schneider is Professor of Religious Studies at Vanderbilt University. Her books include *Beyond Monotheism: A Theology of Multiplicity* and *Re-Imagining the Divine: Confronting the Backlash Against Feminist Theology* . She has contributed to multiple anthologies and journals, co-written *Awake to the Moment: An Introduction to Theology* with Stephen G. Ray and co-edited *Polydoxy: Theologies of Multiplicity and Relation* with Catherine Keller. She continues to work on decolonial, feminist queer poetics in the construction of theological ideas, particularly in conversation with philosophical systems beyond the European canon. She has also been active in the American Academy of Religion, completing two terms on the Committee for the Status of LGBTIQ Persons in the Profession.

Lisa Stenmark teaches Humanities and Comparative Religious Studies at San Jose State University. She is the author of *Religion, Science and Democracy: A Disputational Friendship*, on scientific and religious authority in public life, and is active in the American Academy of Religion and the Arendt Circle. She has written numerous articles on the relationship between religion and science particularly in the context of democratic discourse and within the broader culture. She earned an MDiv/MA from Pacific Lutheran Theological Seminary/ the Graduate Theological Union, and a Ph.D. in Religious Studies from Vanderbilt University. In her spare time she trains for triathlons, practices Aikido, and reads too much science fiction.

Emilie M. Townes is the Dean and the E. Rhodes and Leona B. Carpenter Professor of Womanist Ethics and Society at Vanderbilt University Divinity School. She is the author of three co-edited volumes and four books. The most recent of her books is the ground-breaking *Womanist Ethics and the Cultural Production of Evil*.

Carol Wayne White is Professor of Philosophy of Religion at Bucknell University, and the author of *Poststructuralism, Feminism, and Religion: Triangulating Positions* (2002); *The Legacy of Anne Conway (1631–70): Reverberations from a Mystical Naturalism* (2009); and *Black Lives and Sacred Humanity: Toward an African American Religious Naturalism* (2016). White has also published articles on process philosophy, religious naturalism, and critical theory. She is currently writing a new book that explores the tenets of deep ecology and insights of religious naturalism expressed in contemporary American nature poets and writers.

Zairong Xiang is postdoctoral researcher of the DFG Research Training Group "Minor Cosmopolitanisms" at Potsdam University. He received his Ph.D. in Comparative Literature from Eberhard Karls Universität Tübingen and Université de Perpignan Via Domitia. His research intersects feminisms and queer theories, philosophical and religious inquiries, literary and visual studies in their decolonial variants in Spanish, English, Chinese, French, and Nahuatl. His book *Queer Ancient Ways: A Decolonial Exploration*, a critique of modern/colonial reception of Mesopotamian and Mesoamerican creation myths is forthcoming with Punctum Books.